Perspectives on
Radio Frequency Identification

What is it, Where is it going, Should I be Involved?

RON AMES

VAN NOSTRAND REINHOLD

New York

Copyright © 1990 by Van Nostrand Reinhold

ISBN 0-442-00406-0

Printed in the United States of America

Van Nostrand Reinhold
115 Fifth Avenue
New York, New York 10003

Van Nostrand Reinhold International Company Limited
11 New Fetter Lane
London EC4P 4EE, England

Van Nostrand Reinhold
480 La Trobe Street
Melbourne, Victoria 3000, Australia

Nelson Canada
1120 Birchmount Road
Scarborough, Ontario M1K 5G4, Canada

Purpose of this Book

Many of you have recently become aware of RF/ID for the first time. Others have been aware of it for some time but have a need to get up to date on this very interesting technology. Or you may want to understand the state of the development of this promising industry and/or of the markets and applications for the technology.

This book looks at these questions through the eyes of several people connected with the industry, through the vehicle of an anthology of articles written for publications dedicated to this field and in papers which have recently been presented in seminars related to this subject.

The author of this book, Ron Ames, attempts to provide context, transitions, interpretations and explanations, from his personal perspective as a consultant and practitioner in this field for some years, and as participant and developer of computers, related products and applications for more than twenty five (25) years. He also provides his analysis, additional insight and projections, regarding the technology and the industry, in the later chapters of the book. The authors' contributions in various forms, constitutes, for better or worse, about one third of the total content of this book.

Since some of the authors have attempted, in the limited space available to them, to address the whole scope of technology, multiple applications and a perspective on the whole industry, it is not possible to arrange their writings by individual subject. As a result, portions of the articles will not relate to the title of chapter in which they are located.

The views of the authors of the articles do not always agree with other authors and in some cases the author of this book will disagree with specific data, points of view or conclusions expressed by them.

Acknowledgments

The author wishes to acknowledge and thank the original publishers of documents included in this book, the Automatic Identification Manufacturers (AIM USA), 1326 Freeport Road, Pittsburgh, PA 15238, Edgell Communications, 7500 Old Oak Boulevard, Cleveland OH44130, Helmers Publishing Inc. 174 Concord Street, Peterborough, NH 03458, Marking Devices Publishing Company Inc., 2640 North Halsted Street, Chicago, IL 60614, and Penton Publishing, 1100 Superior Avenue, Cleveland, OH 44114, who granted permission to publish these articles in the hope that it would further the development of the RF/ID industry, and the authors who contributed their knowledge and insight in the original articles.

Also, thanks to my partner and friend, Rich Pollack who has contributed greatly to my understanding of this field, and to my wife, Marilyn and my daughter, Tamara who contributed typing, editing and other skills.

Thanks also, to Judith Fowler of ApplePie Communications whose desktop publishing skills and contribution to my computer literacy are greatly appreciated.

And finally, I thank my Mac Plus. Without it, this wouldn't exist.

Introduction

The first chapter of this book contains articles which introduce the reader to the more general field of Automatic Identification (Auto ID) and show how RF/ID relates to other technologies in that field. It is easily possible that you may conclude that some of the products which can be classified as RF/ID may better be considered as part of other industries. Some method of classification is useful until your level of understanding is sufficiently broad and deep that you can reclassify them in ways which better suit your specific purposes. You will find increasing numbers of sub-classifications in ensuing chapters, particularly in Chapter 3.

Whatever your reason for interest in this field, you would probably like to know something about the size of the industry, is it growing etc. Chapter 2 provides some insight into the trends. Articles in Chapter 5 also address this subject.

Though certainly not exhaustive, Chapter 3 includes discussions of the technologies used; how products are categorized technologically; and why specific capabilities are designed into products. The last chapter will indicate additional sources of information on the technologies employed in this field.

Currently successful applications of the products in specific installations are addressed in Chapter 4. These are by no means comprehensive but are representative examples which can provide a springboard for creative thinking about potential applications from the readers' experience.

The future of RF/ID is addressed in Chapter 5, as well as some issues facing the industry. Several market opportunities, which appear to be within the near term capabilities of the technology, are identified.

In Chapter 6, the broad question of what is needed, in respect to several factors affecting the industry, in order for it to reach it's full potential in the longer term, is discussed. Some potential strategies are offered for you readers to achieve your objectives. Whether you are planning to become a new supplier, are currently a supplier, an O.E.M. customer, a system integrator or an end use customer of the industry, some things that you may want to consider in setting your strategies, are presented.

Chapter 7 offers a summary of the author's conclusions about the industry and suggests other sources of information and education on the subject.

The authors' credits appearing after the "Reprinted by permission.. " statement   at the end of articles, are generally as they were in the original publication.

Finally, a Glossary of Terms is provided at the back of the book as a reference aid as you read.

Let's get started, shall we?

Contents

What Is Auto ID?

and how does RF/ID relate to other Auto ID technologies?

These questions are addressed by the articles in this chapter which give you a small taste of several different techniques used to provide the identification and often other information, as well, about entities which may be alive, such as people or animals; or not, such as components used in manufacturing or finished goods, such as an automobile.

The term "Auto ID" is much too narrow and restrictive to include the entire scope of any technique called by the name. But, automatic identification is almost always one of the functions being performed by these subsystems, since any other information provided which was not related to a specific entity or context would be worthless.

Comment: 1.1

The following article explains the need for automatic identification (Auto ID), in general, and provides a brief description of the technologies which are most often included in the definition of Auto ID. Other technologies, not addressed in this article, which are are also normally included, are vision systems, magnetically coded ink (MICR) used on checks, magnetic stripe used on credit cards, and Wiegand wires and barium ferrite inserts used in access control badges or cards.

1.1 Run with the Best

An Overview of Automatic Identification Technologies

Alan E. Gold—ID Systems—© April, 1988

New technologies mean old error-ridden entry problems solved, and new risks ventured. The best way to risk-proof your system is to understand the technologies first—how they operate, where they work best, and where they don't work. Here's a basic guide:

Bar Code Systems

Bar codes are a series of thin or thick lines and spaces that signify numeric, alphabetic, and control characters; their success lies in their relative simplicity: reading devices detect either the existence or absence of a bar —or space— and transfer that information to a processor —be it a handheld data collector or a PC— for decoding and processing. The likelihood of gathering incorrect data from such a system can be as small as 1 in 1.2 million characters read, a far cry from error rates found in human data entry.

A typical bar code system consists of labels, scanner (reader), decoder, and processor.

Bar code labels have to be printed quite carefully and exactly, since bar code scanners and decoders are not very forgiving. Technology today has produced dozens of dot matrix, laser, ink jet, and full impact printers that perform as required by a bar code system.

Bar code scanners come in two broad categories: contact and noncontact. The most common, and typically least expensive, option is the contact wand. An LED sensor inside the wand reads bar codes it is passed across by detecting the relative reflectivity; it passes on that information for decoding and processing. The wand's advantage is its high performance to cost ratio. As a contact device that also depends on operator skill to manually pass the wand over the bar codes, it may not always be the best choice of scanner. In a relatively low throughput, comfortable environment, a contact wand is completely satisfactory.

The noncontact scanner—be it laser, LED, or CCD (charge-coupled device)—although substantially more expensive, has significantly greater throughput and flexibility, and provides practical automation in environments where it is difficult, dangerous, or impossible to get close enough to touch the bar code label. Operating distances range from a few inches (CCDs) to several feet (lasers). Most devices of this type have a depth of field that allows for human fallibility or unexpected distance variations.

Often a part of the scanner subsystem, the decoder translates the information gathered by the scanner into information that is usable by the rest of the system. Here the "decode algorithms"—a computer program—turn the scanner's output into "normal" computer data for processing. These decode algorithms, more than the scanners, are responsible for the accuracy of a bar code system. Most all commercially available bar code systems now operate with relative flawlessness.

The processor can be a personal computer, a full-blown mainframe, or a handheld data collection device. All have the ability to process bar code data.

Handheld data collection systems enable on-the-spot processing or verification of scanned data, storage of that data for later use, or remote, real-time processing with radio-linked transmitters. Bar code systems are not appropriate in all environments. They are most useful when used to gather a large number of small pieces of data. The strength of bar codes is in its use in counting, tracking, and identifying applications, typified by the supermarket checkout line. If large, descriptive pieces of information are to be gathered, other technologies are more useful.

Optical Character Recognition Systems

Optical character recognition (OCR) technology, most widely implemented via handheld wands that read single lines of alphanumeric information, use a matrix of detectors located within the wand to "read" reflected images of the scanned data. The decoder attached to the wand makes a "best guess" as to what characters it is seeing. This information is passed on for storage and/or processing.

OCR technology has some problems that have shifted its growth in application areas. OCR is primarily a contact technology; to read OCR characters you have to brush the wand over them. Because of the relative sensitivity of OCR decoding systems, OCR printing must adhere to fairly stringent guidelines to be successfully scanned. Quality printing has been hard to come by, and in high volumes, difficult to maintain. Although accuracy is high—though not as statistically impressive as bar code—the "first read rate" is substantially lower. That means more dependence on operator performance, which is difficult to control in many environments.

OCR page readers, devices that can "read" and transfer typed or printed manuscripts, bills, letters, checks, and such into usable computer data, have hastened automation in a number of environments. Law firms are now able to add years of legal information and records into their word processing systems for later use. Libraries can save deteriorating materials, and publishing houses can go from typed manuscript to finished book without a typist.

Few, if any, systems can indiscriminately gobble up all words spread before them. Most OCR page readers require that documents contain certain character fonts and spacing. The more sophisticated units can differentiate between several fonts and sizes at one time, but with lowered accuracy and an increased price. Speed is also a function of sophistication and cost. Advanced systems can process thousands of pages in a short time.

Speech Recognition Systems

Speech recognition is distinct from speech synthesis and voice store-and-forward systems in that it actually "hears" words and either performs predesignated tasks based on those words or stores the spoken information for later use. Existing technology generally requires single, discrete word entry, though a number of companies offer systems that can recognize limited continuous (natural) speech. Systems that require the user to "train" the unit to his or her voice—speaker dependent—or to recognize spoken words form a large number of users—speaker independent—are available. Both systems use a pattern matching process to determine the identity of a word, but in a one user/one system environment, the speaker dependent system's matching pattern is made up of the user's own voice characteristics, while a speaker independent system depends on a database of "averaged" patterns drawn from a large sample of voices. Not surprisingly, the speaker dependent system is more accurate; speaker independent devices provide greater flexibility.

A typical speech recognition system uses a handset and microphone to enter words into the computer-based system for processing. Wireless microphone units are particularly useful in environments with dust, dirt, temperature extremes, and other hazards. Speech synthesis devices are often used in conjunction with the speech recognition unit to provide feedback and direction to the user. These systems provide automation capabilities in environments where users are not computer literate or where keyboards are similar input devices are not practical.

RF Identification Systems

Often confused with radio transmission devices used in remote bar code and other portable applications, RF (radio frequency) identification systems are different in both design and use. An RF identification system typically consists of reusable, programmable tags placed on items to be tracked and a reader/receiver that captures the information contained within the tag when it is passed within range of the receiver. Tags may contain as much as 32,000 bytes of information about the item. In some cases the data can be altered by the receiver, thus providing a continuous update capability.

RF tags are essentially tiny computers embedded in a small container sealed against contamination and damage. Some contain batteries to power their transmission; others rely on the signal generated by the receiver for the power necessary to respond to the receiver's inquiry for information. The receiver is a computer-controlled radio device that captures the tag's data and forwards it to the host computer.

In a typical warehouse installation, travelling bins filled with items to be distributed among numerous trucks are identified and routed to the appropriate location through the use of RF tags and automatic transport equipment. Accuracy approaches 100 percent. This application can eliminate the need for human interference in the sorting and moving processes, and speed up throughput considerably.

Other uses of RF technology include truck and trailer tracking and wildlife identification. Low-frequency RF tags can be read from and written to over distances ranging from a few inches to several feet; microwave-based devices are capable of distances of 20 feet or more—though the power requirements and microwave radiation emitted from such systems requires special consideration. Tags, capable of storing entire shipping manifests, are extremely useful in minimizing access and processing time on a firm's computer system: paperwork can be generated on site without data processing backlogs.

Smart Cards

Take a microprocessor, add some memory, package it into a slightly larger than normal plastic credit card and you've got smart card technology. Used mainly in banking and security environments, smart cards provide protection against fraudulent or unauthorized use that far surpasses traditionally encoded credit or pass cards. The cards can even "defend themselves" against counterfeiting or unlawful use by becoming inoperative if a password is not correctly entered.

Smart cards can contain generic data about its owner, eliminating the need for multiple credit, telephone, and bank cards. Information such as emergency health data can be incorporated into the card, and in some cases frequently used telephone numbers can be stored, thus providing a pocket database for the owner.

Smart cards have made a greater impression in Europe than in the United States primarily due to the prior existence of an extremely sophisticated credit network across this country. U.S. credit companies have recently taken longer looks at the technology in order to keep up with the increasing demands of the credit environment and to offer value-added capabilities to their subscribers.

Security environments have begun to look at smart card technology as a way to tighten control over access to sensitive or restricted ares. Nor have industrial applications for smart cards been forgotten. Several hardware manufacturers use smart cards to provide "personality" data to their hardware devices. A printer's font can be changed simply by inserting a smart card containing the control data for the printer. Application programs of limited length can be encoded in a smart card providing limited functional access to users without the risk of hackers overriding security measures. Applications will continue to emerge for this young technology.

Touchscreen Systems

The basic premise behind touchscreen technology is that it is easier to touch a spot on a display screen than it is to enter a command on a keyboard. Where computer literacy is not high—in a restaurant, for example—but automation is important, a well-designed touchscreen system is an outstanding data entry solution.

Touchscreen systems use several different methods. Some systems surround the display with rows of infrared LEDs, effectively turning the screen into an invisible grid. When something enters that grid, the location is passed to the system for action.

More recently, touchscreens have been developed that use the capacitance of the human finger to activate a grid contained within a clear membrane attached to the front of the screen. In a manner similar to those "magic" lamps that adjust their intensity and turn on or off with a simple touch of a finger, these touchscreens translate the touch of a finger into a location on the matrix which is then passed to the computer.

Touch screen subsystems are available, but most end user applications for this technology incorporate the screen into a larger system designed for a special purpose. Grocery and department stores have begun to use such systems to assist customers in locating departments and providing information that used to be available only from a manger or an information booth. These systems keep their computers out of the way, presenting only a color monitor in a pedestal display to the user. Colorful graphics guide the customer through various menus and options, providing easy-to-read blocks to touch to select that item.

Likewise, restaurants successfully use touchscreen systems to enter orders, tally meal checks, and speed customer orders. Little or no training is involved, and the same sophisticated graphics that are available in today's software can be used to guide the user to painlessly through a complicated transaction.

Touchscreen systems for manufacturing use icons to guide computer-illiterate employees through manufacturing procedures while collection production data.

Though questions of durability and overall technology usefulness persist, users of these systems—they are becoming more numerous every day—find that these systems are not stressful to use and are straightforward to operate, with throughput equal to traditional keyboard entry. Whether functioning as combination displays and keypads in point-of-sale terminals, as an on-line directory in a hotel lobby, or as an interactive learning system for children, touchscreen input provides many space, performance, and human factors advantages.

The automatic identification systems described here, different though they may be in technology or application, share a common characteristic: each is designed to provide automated, accurate, and timely collection and processing of information. Keyless data entry is a logical progression in the continued computerization of business and industry.

Some may argue that all this automation is a double-edged sword, trapping its users even as it attempts to free them from drudgery. But for the growth of auto ID technologies, most of us are quite willing to take that chance. The benefits from such automation seem to be well worth the risk.

Alan Gold is vice president of marketing and sales for Digitronics, Comtec Information Systems, Inc.

Comment: 1.2

The next article provides advice and guidance to those responsible for making a factory productive. It is more broadly applicable than that. There are however, a few specific points I feel I should draw to the readers attention.

I believe most practitioners in RF/ID would challenge some of the authors ratings of RF/ID in Table 1-1.

Specifically, the tagging reliability would be rated as excellent, because the tags have to be read by a reader to confirm their ID. Confirmation is visual as it is in bar codes and text recognition. I would have expected voice recognition to have a poor rating.

Data security is not defined, but in this context appears to mean data integrity or freedom from reporting erroneous ID's, rather than ability to prevent unauthorized persons from reading data from the tags. RF/ID normally is rated very high since most products have rather elaborate error detection schemes. The only acceptable failure mode in most products is, if the data from more than one reading —sometimes several—are not identical or if parity or check sums are in error, no data is reported to the host.

The table also indicates that portable readers are not available. In fact, some vendors do offer portable readers.

The author states that "a high-frequency (GHz range) base station scanner" is used. Very few systems' readers broadcast in this frequency range. A few systems broadcast at 910-925 MHz (0.915-0.925 GHz), to the tag, and receive 1.812 -1.850 GHz from the tag. Most current systems use frequencies in the under 10 MHz range.

The author uses the terms active, passive and locally-powered. The definitions he uses for these terms are those promoted by a single manufacturer. They are defined quite differently than the definitions published by the Automatic Identification Manufacturers (AIM) trade association and used by other manufacturers. The terms used by the industry which are equivalent to these are; passive = Surface Acoustic Wave (SAW) or multiple tuned circuits; active = passive; and locally-powered = active. I know this is confusing, however, these terms are more thoroughly discussed in Chapter 2.and in the Glossary.

1.2 THE ID FACTORY

Automatic Identification Improves The Bottom Line

Kevin R. Sharp—ID Systems—© June, 1987

Every manufacturing manager needs solid, consistent data about his manufacturing operation so he can make sound judgments about how to improve it. Whether the goal is to increase production capacity, decrease costs, or improve quality, the first step is to accurately monitor the current manufacturing process. An automatic identification system will improve data collection precision while decreasing the time spent by direct labor personnel to gather the information management needs, freeing them to build product.

Automatic identification is well suited to the wide variety of repetitive, time-consuming operations in a manufacturing environment. Many of these operations are probably already done manually. One easy application is time and attendance tracking. Most manufacturing concerns require each operator to record the time spent on every operation performed during the day. Since the operator is usually required to make each day's entries total eight hours, to save effort the times entered are usually rounded off to the nearest half hour. If an automatic system is installed to track the beginning and end of each operation, and it is easy and convenient for the operator to use, suddenly management will discover that each operator has performed not 10 or 15 tasks each day, but 20 or 30. In this time of decreasing profit margins and increasing foreign competition, a manual labor-tracking system accurate to only plus or minus one-half hour puts a company at a dangerous disadvantage.

Focus on a Need

Once you're determined to install an automatic identification system, focus on the results desired, not on the available technologies. Prospective users are easily caught up in making sure they get the latest technology, spending too little time thinking about the human factors required to make the upgrade a success. Pick a specific goal, write down the results you want to achieve, and only then start looking for the best —not necessarily the newest—technical solution. Goals should be phrased in terms of what data is needed and how the data should be presented. For example, here are some realistic goals and some goals that may present problems. Note that realistic goals for a manufacturing automatic identification system usually start with the word "identify" not "buy," "install," or "find."

Good Goals

1. Identify all labor expenditures down to the minute.

2. Identify all efforts of direct labor that do not directly contribute to the production of items for sale.

3. Identify all workstations that produce more defects than the plant average.

4. Identify all workstations that consistently have idle time.

5. Identify all material shortages as manufacturing kits arrive on the floor.

Questionable Goals

1. Buy the newest remote identification technology available so that we don't have to buy another system for a long time.

2. Find the fastest communication network to use for factory data collection.

3. Install a terminal on everybody's desk.

After deciding what data needs to be acquired, the next task is to determine how to collect it. Again, focus on your operation: where are the best places to pick up the information? If the goal is to track materials, there should be a label that moves with the articles and it should be read at each critical point in the plant—such as when it passes incoming inspection, when it goes into the stockroom, when it leaves stock and arrives on the manufacturing floor, and so on. If the goal is to track labor, each operator should carry a label or be assigned an ID code, and each workstation should have a menu of labor-ticket options from which the operator may choose . If the manager focuses on the needs of the people using the identification system, and considers when they will be expected to enter information, the choice of an identification technique will be much easier.

Choose a Technique

The goal of all automatic identification systems is the same: get information correctly into digital form for use by a computer system without requiring an operator to enter the data from a keyboard. The manager assigned the task of tightening controls over material or people, either for just-in-time manufacturing operations, or confidential data access control, will find a wide variety of techniques available to solve the problem. The real task is to decide which technique best suits the application. Table 1-1. is a summary that will help the manager in this task.

Table 1-1. Automatic Identification Techniques Summary.

	Magnetic Stripes	Bar Codes	RF Tags	Text Recognition	Voice Recognition
Scanning Distance	Contact	Contact to 36"	Up To At Least 60"	Contact	<36"
Industry Standard	Yes	Yes	No	No	Not Applicable
Handheld Scanners	No	Yes	No	Yes	Yes (Radio Links)
Tag Making	Special Device	Easy	Special Device	Easy	Not Applicable
Tagging Reliability	Poor	Excellent	Fair	Excellent	Excellent
Data Security	Good	Good to Fair	Good to Fair	Fair to Poor	Good to Fair

All automatic identification systems consist of an identification tag and a scanner used to identify the tag and report its presence to the host computer. To track the location of people or materials, the tags will be attached to the object to be tracked; labor ticket reporting usually requires a preprinted menu of ID tags that the operator scans to report the initiating or completion of a task, as well. In addition to tags and scanners, there must be some method to make the tags. Sometimes this capability can reside at the user's installation; other times the tags must be specially made by an equipment manufacturer.

To determine which automatic identification technique will work best in a given situation, six things must be evaluated: scanning stations, tag marking, environmental limits, data security, reliability of tagging, and network compatibility. The question of tagging reliability is often overlooked in the design of an identification installation. Remember that the best automatic identification equipment in the world won't work if the wrong tag winds up on an inventory bin. Expansion also should be considered—all the identification techniques and workstations needed now and in the future should be able to function together in a smoothly operating network.

What Is Auto ID?

The Most Versatile Manufacturing Input

A bar code is a printed pattern of bars and spaces, commonly seen on consumer items in the grocery store. Some bars and spaces are thick and some are thin, and in this way convey binary information to a bar code scanner designed to read them. Several bar code symbologies are available for different purposes, some designed to convey the most information in the least amount of space, some designed to represent all the characters on a standard computer keyboard, some designed to be easily printed by inexpensive printing techniques, and some designed to be read with great reliability. The best type of bar code to use in a manufacturing environment depends on the application, and can best be determined by talking to the identification system vendor.Generally, bar codes are the most flexible and least expensive identification and data entry technique available to the manufacturing professional.

Bar code applications in manufacturing fit into two main categories: menu selection and identification. Menu selection effactually lets an operator fill out electronic forms by passing a bar code wand—about the size and weight of an expensive mechanical pencil—across different bar code patterns preprinted on a menu available at each workstation. This is often the first step a manufacturing operation will undertake in automatic identification technology. The other common use of bar codes in manufacturing is for material or personnel tracking, where each unit of material or employee is affixed with a bar code label which is read at critical locations, automatically tracking the progress of a manufacturing kit or locating key people.

Incorporating a bar code menu selection system is easy if the focus remains on the real worth of automatic identification: quick and accurate entry of critical data. Again consider an assembly operation that requires its direct labor to write up labor tickets accounting for each payroll hour. In the interest of saving time, looking good to management, or simply laziness, each operator will round off the time spent on each task to the nearest half hour or the nearest hour.

If, however, a small, non-threatening terminal with bar code input is available and a menu is provided, the operator will spend less time providing this critical data and the information will be accurate to the minute—since the time of each data entry is automatically tagged by the transaction processor or network controller. Minimizing operator entry time and increasing data precision is the key to automatic identification profitability to the manufacturing company.

Bar codes are scanned by passing a beam of light across the label. The reflection is analyzed to determine the content of the label. An operator may physically pass a wand across the label or the scanner can create a moving beam. The pistol-shaped laser guns and supermarket checkout scanners are moving beam scanners. Increased competition and new developments in solid-state laser physics will drastically reduce the cost of moving beam scanners and make them less easily damaged when dropped.

A development finding its way into fixed station bar code readers—for use on conveyor belts, for instance—is to view a bar code with a charge-coupled-device (CCD) camera, like those sold for use with VCRs for home movies. These cameras, with sophisticated signal processing techniques, result in a scanner that is less affected by dirt and by tag orientation. The cost of these devices is high compared to bar code wands and moving beam scanners, but if an application calls for a large depth of field—the difference between the nearest and farthest bar code that can be read—they may be the only solution.

Magnetic Stripes

The chief advantage of magnetic stripe identification is also its chief disadvantage: it is difficult to copy magnetic stripe identification tags. If your application calls for security in the transmission of confidential data, then magnetic stripes are a good option—virtually all credit cards and automatic teller cards use magnetic stripe technology.

There is an industry-wide standard for magnetic stripe data encoding. Therefore, cards coded on a machine made by one manufacturer can be read on any machine conforming to the standard—necessary in the financial community. On each stripe are three low-density tracks of data: tracks one and three record binary information at 210 bits per inch; on track two, the most commonly used track, data is stored at 70 bpi. If your application requires you to store lots of data, magnetic stripes aren't for you.

The magnetic stripe scanning stations, usually called card readers, require good alignment between reader and stripe. The card readers must have some sort of mechanical alignment assistance, effectively eliminating handheld magnetic scanners. Since you can't hold the scanner, you can't affix a magnetic stripe to a box, but would have to, say, put the card in a pouch on a box. To identify the box, an operator would remove the card and insert it into a card reader. This could cause identification errors if cards got mixed up, since the identification tag is not effectively attached to the object being tracked.

The final two features to consider about magnetic stripe marking are tag production and environmental limits. Magnetic stripe data must be encoded on specially-made cards by machines made for the purpose.

That specialized equipment is required is an advantage in applications where security is a consideration, such as financial industries or personnel access control.

Because magnetic media is susceptible to strong electromagnetic fields, care must be taken to protect the cards. Problems have been experienced around the newer machining tools that use intense RF energy to shape plastics. Credit cards have been ruined around these machines.

RF Tagging

Current techniques for reading bar codes require that the scanner be no more than a few feet from the bar code, and usually require that the bar code be generally oriented perpendicular to the scanner's beam. For applications where it is necessary to identify objects at further distances or with less precision to orientation, particularly applicable to personnel identification, radio frequency identification—RF tagging—is available. This technique uses a high-frequency—GHz region—radio base station scanner to detect mobile radio frequency tags. A major manufacturing application of RF tagging technology is personnel tracking and access control in sensitive or classified areas. Unlike using bar code or magnetic stripe for personnel identification, where the person must deliberately insert a badge into a reader to be identified, an RF tag will be read whenever the wearer passes a reading station, even if the wearer tries to hide the badge from the reader.

All RF tag systems work fundamentally the same. A base station transmits a signal to any tag within range. The tag replies, usually on a second frequency, and in some way identifies itself. Each manufacturer involved in this business uses slightly different techniques and protocols; therefore, the selection of a vendor ties the user into continuing to buy product from the same manufacturer. The RF tagging companies desperately need to develop industry-wide standards if they are to achieve anywhere near the widespread acceptance of bar codes or magnetic stripes.

Three basic tag styles are used by RF tagging companies: passive, active, and locally powered. Totally passive RF tags receive the incoming signal, distort it in some recognizable way, echoing some fraction of the incoming signal back to the scanner. These tags are least affected by environmental conditions, but the tag ID is a function of the tag manufacture, dictated by the etching patterns in the circuit chip. Therefore, if the user wants a tag with ID number 1234, for instance, he must order a tag with that number built-in.

Active and locally powered RF tags can be programmed by the user to transmit whatever ID number or other information is required back to the scanner. Both are low-powered integrated circuits with a transmitter, a receiver, and some memory. When they receive a signal from the base scanner, they transmit a signal determined by the contents of their memory. The only fundamental difference between passive and local powered tags is the source of their power. Active tags actually derive their power from the energy in the scanner's signal, while locally powered units contain a battery. In practice, locally powered tags can contain much more data than active tags. Complete manufacturing information can be contained on some of the locally powered tags, rather than an index code to the information that is stored in some remote database.

Any RF tagging system is subject to radio interference, from outside sources and from multiple tags responding simultaneously. RF tags are best used in applications requiring identification of moving objects separated by a reasonable distance. The technique should not be considered for applications requiring the identification of a particular object close to many other RF-tagged items.

Most RF tag companies do not offer a full line of automatic identification techniques. However, most of the scanning stations are available with standard RS-232C serial communication interfaces that can be hooked up to an integrated data collection network.

Text Recognition

A desirable feature of automatic identification techniques is that the code be in human-readable form. If a human can read the tag to which the computer will respond, it is easier to detect tagging errors before they become bad shipments or wrongly manufactured products. One identification technique that has failed to live up to its promise in this area is automatic text recognition. Commonly called optical character recognition (OCR), this technique has found retail applications in some large department store chains, but is plagued by a poor read rate. Next time you are in a store that uses this technique, watch how many times the check-out person has to scan a code before the cash register responds.

OCR scanning techniques fall short because they are very sensitive to the orientation of the characters in respect to the scanner, and they require good contrast between the character and background (no dirt). Both requirements severely restrict the techniques' industrial and retail applications, but are not a real problem for full-sheet document readers. In these readers, full-sized typewritten or printed sheets can be fed through a device that provides the mechanical alignment necessary for good response, and the document is read into a computer system.

The same CCD and signal processing techniques mentioned for bar code scanners may find applications in OCR, but I have not seen any products currently available using this technique.

Voice Recognition and Speech Synthesis

Although not technically an automatic identification technique, since it requires a human operator, voice recognition systems offer an excellent data input technique in applications where an operator is present irregardless (sic) of the entry technique. Excellent applications include data entry at microscope-based inspections, package loading stations, and other places where an operator is performing one function that requires human participation and another that is purely data entry function, particularly if the non-data-entry task keeps the operator's hands and eyes busy. Think of voice entry systems as an automatic identification system in which the human is the vision system and processor, and the voice recognition station is the communication device.

Two categories of voice recognition equipment are available for industrial use: isolated word recognition and connected speech systems. Isolated speech systems require that the operator separate each word to be recognized by silence, while continuous speech systems allow the user to speak in full phrases or sentences. Either technique lets context clues be built into the vocabulary to enhance recognition performance.

The cost of voice recognition stations has dramatically decreased in the past few years. Systems vastly superior to the older, $20,000 units are currently available for less than $6000, and the substantial user programming previously necessary is no longer required. Look for voice recognition to be a common data entry technique before the end of the next decade.

The Personal Touch

One extremely critical aspect of integrating an automatic identification system into a manufacturing environment that is often overlooked is the perception that ultimate users—the production operators—have of the system. The operators must feel that the system is designed to make their job easier and more enjoyable, not that automation is going to put them out of work or turn their supervisor into an all-seeing Big Brother. Two ingredients necessary for user acceptance are good management/employee communication and benign-looking products.

As soon as the management of a manufacturing operation decides to install an automatic identification system, or better yet, even while management is discussing whether to install one, that is the time to start talking to the workers. The communication can take the form of formal face-to-face meetings, newsletters, or just plain talking on the manufacturing floor or in the break rooms. Management should focus on those unpopular day-to-day tasks that the ID system will make easier or eliminate—especially how it will decrease the amount of paperwork required from each employee. Remember—no automation system, no matter how perfectly designed and installed, will live up to its promise if the operators don't want it to work or don't believe it will help them. Incidentally, a properly designed and installed automatic ID system should decrease the paperwork load. If it doesn't, serious thought should be given to why it doesn't.

Another aspect that will increase the operator's acceptance of a new automation system is the look of the products they will be using. High-tech polished aluminum may look slick at trade shows and may impress the heck out of the management information systems people, but round edges, soft keys, and earth tones are much less threatening to the operators on the production line. Save the Star Wars panels for the stereo system in the break room.

Beyond Hardware

After factory data has been put into computer language by an automatic identification system, the job is only half done. The raw data must be manipulated and displayed in such a way that it allows the manufacturing managers to easily see what is going on in the plant. This manipulation is done with software, a set of computer programs responsible for accessing the data provided by the automatic ID system and presenting it in an understandable way. Many software packages are available that allow the user to automatically track time and attendance, material control, shipping coordination, and many other manufacturing-related tasks. If you are interested, go to an automatic identification trade show or two, or talk to some hardware vendors. The more established hardware companies, like Burr-Brown, have developed software packages for their systems or have developed relationships with software companies to support specific manufacturing needs.

Kevin Sharp is an engineer with Burr-Brown Data Acquisition and Control Systems division in Tucson, Arizona. He is currently researching alternative techniques for automatic identification systems.

Comment: 1.3

The following article is intended for an audience interested in material handling in factory environments. However it provides an excellent introduction to RF/ID and positions it relative to other technologies commonly used in an industrial environment. In the Basics section, the author states "Low frequency systems operate in the radio or TV broadcast range or 100 to 600 kHz." The word "in" should be changed to "below" and "600" should be changed to "400."

1.3 RF/ID

electronic codes for automatic identification

Gene Schwind—Material Handling Engineering—© September, 1987

Here's how radio frequency identification works and how it's being applied in industry and commerce.Opportunities for RF/ID to add value to your material handling system abound.

Radio frequency identification (RF/ID) is a method of coding just about anything that moves with invisible information that can be read later or used to cause an action for some purpose. It is a means of automatic identification.

The technology consists of tags and readers which are analogous to bar codes and scanners used with more visible marking.

What makes RF/ID different

RF/ID is emerging as a separate automatic identification technology where:

- The code can be made invisible.
- The code is difficult to counterfeit.
- The code is highly tolerant of abuse.
- The code can be made environment and weatherproof.
- The code must contain more diverse information than is available with other code forms.
- Coded information must be changed on the fly.
- Outside power is not available for signaling.

Applications abound. RF/ID can be used to code animals, places and things. It can be placed where no other code forms can be considered. It has a long life and almost endless numbering capability. Consider these:

• People, livestock, laboratory animals, fish, and many other live species fit the animal category. A photo ID badge permits a security guard to ascertain identity but a hidden RF/ID code in the same badge is needed to open gates, doors, or computer files selectively to the employee.

• Livestock may be coded with a collar and code tag that could be used to record their movements and allot feed or access to it. The tag can also sort livestock for breeding.

• Laboratory mice all look alike but an injectable code transponder serializes each to permit sorting if a cage is left open and to accurately record experiments.

• Place or positions are important to many operations. Guided vehicles can use RF/ID to locate pick-up and drop-off points. Pneumatic tube carriers actuate switches or announce position or control traffic using individual hidden RF/ID code tags.

• Place or position can be identified as a check, demarcation, action or identification point.

• Things make up the largest potential group of possibilities for RF/ID.

Whole vehicles or components of them can be individually identified. The fixtures that carry assemblies through processes, the containers that hold piece parts or individual parts or assemblies themselves can be identified using RF/ID.

Containers from tote box size to ISO intermodel size as well as the chassis and trucks that transport them can carry RF/ID. The hidden tags on vehicles can open gates, direct a particular vehicle to a specific loading dock, record tolls, actuate refueling pumps and generate records.

An area for increasing activity for RF/ID is in production tooling. Tools, tool holders, jigs and fixtures contain discrete code buttons to keep track of tools, to schedule them between machines, to track and anticipate tool wear, and to account for expensive, contract-captive jigs and fixtures. These tags are impervious to vibration, coolant, or chips.

Basics of the system

There are two types of systems: low frequency and high frequency. This refers to the wave length of the signal used to activate the code carrier or tag.

Low frequency systems operate in the radio or TV broadcasting range or from about 100 to 600 kHz.

High frequency systems are in the microwave range and typically from 900 to 1000 MHz.

Each of these systems operates differently, uses different components and has some capabilities that are the same and some being advanced. High frequency systems have somewhat broader capabilities such as read range.

Code tags

The differences in tag sizes, complexity, how they are powered, and how they are read may vary from system to system. The more powerful the tag and reader technology, the faster the information can be transmitted or received.

Another way of looking at it is that fast moving tags or those to be read at some distance—up to 40 feet—from the reader are available in high frequency or microwave system. Moderately fast moving tags or applications where the tag will pass close to the reader are usually low frequency systems.

The tags themselves may be in the shape of a tag, block, credit card, button, cylinder or almost any other form and size. They may be encapsulated, laminated, and embedded in protective materials as well as enclosed in a variety of housing shapes including glass tubes.

A tag may simply contain a code or serial number. More complex tags may contain more information. High and low frequency tags can contain over 128,000 bits of data.

Tags may be active or passive. This means that they have a built-in battery that can send back information faster or over a longer distance because they have their own power source to boost the range.

Passive tags have no battery or on-board power source. The power needed for the tag to identify itself or provide encoded information comes form the reader or interrogator that energizes them.

Simple tags can contain a number or simply indicate presence. The smallest of the tags, those that can be injected into live animals, can contain serial numbers into the millions.

Like the number represented by a bar code, tag serial numbers may reference any amount of information in a computer data base. Thus, automobiles could be serially coded and the computer data base could contain all of the background information, from whatever source, pertaining to that tag and serial number.

Some tags, such as those used on clothing to discourage theft from stores, simply signal a presence. The code means nothing until the garment is moved past a reader that might be set up at the door or exit from a department. When the tag is read, it sets off an alarm. The tag is removed at the cash register when a sale is made.

More advanced high and low frequency tags can contain production information or at least much more information than simple serial numbers. In manufacturing operations they can contain machine instructions and sequence data.

The advantage of such tags is that it relieves the main computer of much of the manufacturing direction and greatly reduces wiring and interconnection between the mainframe and remote operations. It keeps local area control resident within work cell areas or machines.

Levels of tag capability

There are basically four levels of tag capacity. For each higher level, both the tag and reader price advance.

- Level 1 is the simplest. It is used for electronic article surveillance or to shut off machinery when someone wearing a tag approaches.

- Level 2 is used for identification and typically provides from 8 to 128 bits of information storage and is used to serially number tags or most any item to which they are attached.

- Level 3 tags typically store up to 512 bits of information and are used to store or record events such as routing, operations, or a history of progress of the item attached to the tag.

- Level 4 tags have the largest capacity capability or over 16,000, 8-bit bytes. They can hold significant amounts of alphanumeric information which is addressable for reading, additions or changes. These tags are, in effect, portable data bases offering interaction without mainframe computer assistance.

A further differentiation in tag types is that some are read only and other are read/write. Read only tags contain information that was encoded at the time the code was manufactured or when the tag was first placed in service. Your bank card has a magnetic stripe with information installed this way.

Read/write tags have the capability of accepting new data whenever they are reused in a process or when some of their information is acted on, causing it to be erased or changed. This tag coding or writing can be done with a special terminal or done on the fly during transfer through a process. Thus, a simple tag on a tote box may contain a customer reorder rather than the permanent tag code being referenced to a customer order programmed into the computer.

Tags with read/write capability offer great potential for all kinds of handling systems.

RF/ID in material handling systems

A tag that has the capability of changing information or passing it on has obvious advantages for material handling systems.

"We got our start in material handling," says Dominick Alston, vice president of marketing for Telsor Corp. "It is an ideal technology for getting information to guided vehicles and automated storage/retrieval systems."

"Our RF/ID does not use SAW technology but what we call electromagnetic inductance. It is a low frequency system but has some advantages over high frequency systems that use SAW technology."

"We can read over-the-road vehicles at 60 mph but we can guarantee 100% good reads at 15 mph, such as when traffic is funnelled through highway toll lanes."

Telsor is taking inventory control seriously. A national automobile rental company is tracking their vehicles with RF/ID. Systems also keep track of country road service vehicles and emergency equipment. In waste management they are combining RF/ID with weight sensors in order to bill customers precisely for refuse hauled and to prevent highway fines for overloads. An extension of the same system is used for fuel management.

Lift truck operation in large warehouses has always been difficult to control. To maximize productivity, getting an accurate picture of lift truck activity is needed. RF/ID offers that kind of "what/when/where" reporting. With the addition of special sensors which add output to that of the tag, records can be produced showing whether the trucks are traveling full or empty.

"There are hundreds of situations where RF/ID can add value," says Alston. "Material handling offers many of them."

In manufacturing at Detroit Diesel Allison Division, for instance, a machine pallet carries an engine block through a range of machining sequences and inspection steps. Initially the engine serial number is written on the pallet RF tag. Every operation performed on the block is recorded by serial number. The data, including guaging, is used for statistical quality control.

At the last station before the block is removed from the machining pallet, the tag information is transferred to a laser engraving machine that converts it to etched identification on the engine block. The information will be placed on a surface machined to receive it.

The blocks and pallets are moved throughout the operation by automatic guided vehicles (AGV's) which also are directed by the tag information.

After machining, the pallet is recycled with new information being added to the tag and a new block casting put in place.

In auto assembly, welding fixtures get a body style number that tells loaders what parts are to be placed in the fixture. As sheet metal parts are loaded into the fixture, the code tag tells each station what body style is being built. The tag will also tell the robotic welders the body style, causing them to change their spotweld programs to fit the style.

In non-synchronous assembly, every auto body or major component can have a slight difference so it is important for the tag to cue the processes on the spot rather than use a communication network. This reduces dependence on the mainframe and permits work to continue even though the mainframe may be down.

In a European auto manufacturing operation, the tag is temporarily attached to the body itself and it will travel through a series of dips and painting as well as assembly. The tag contains information telling stations what special features and accessories to add. With vehicles being produced for a world market, the differences, aside from accessories, can be significant.

Factors to consider

There are many decision points on the way to selecting an RF/ID system. Some system suppliers say that there are few plants or operations that cannot be improved with some level of RF/ID.

But there is more to RF/ID than its consideration as another option. Ron Ames of Ames & Associates suggests you look at these points.

- Cost: first, ongoing, and expansion.
- Capacity: initial and future needs.
- Reliability and accuracy: up-time and percentage of correct reads.
- Range and reliability: how far away read will take place and under what circumstances.
- Programmability: read only or read/write capability.
- Speed: speed of the tag when it passes the reader.
- Life: how long the tag will last.

Trends

If there is a solid trend today, it is toward low frequency systems. Perhaps it is that the low frequency system requires less concern for microwaves in areas of the plant where people will work. Some suppliers indicate that the low frequency system uses lower cost components and circuits that can be reduced to microchip levels. This, they say, indicates that the costs will come down as the chip technology increases. And, there are many more suppliers of low frequency than high frequency.

As many RF/ID systems are local to the operation—RF/ID used only within a machining operation, for example—it is vital that they be reduced to board or component add-on level. The parallel to this is the wand scanner and decoder board that can be added to any personal computer to interpret bar codes.

Another trend is to use RF technology in combination with other advanced communication methods. For instance, new guidance technologies for AGV's may only guide a vehicle from point to point without a wire in the floor; but at decision points, changeable information may be transmitted by RF/ID tags. Fine positioning may also be done by such tags.

"It is true that most of the RF manufacturers are in the low frequency area." says Ron Ames, "but there are some interesting things being done with high frequency too. They also have the potential to get their tag or transponder costs down if high volume users adopt the system."

"Two potential developments that will impact on material handling are the AT&T smart card with a capability of 1000 to 3000 bit storage and read/write capability and the English GEC card with 64K bytes of storage. While a little thicker than a credit card, these may be reduced to the ISO standard thickness of 0.031 inch. When such cards are created in volume, such as would be used for telephone calling, the price will definitely have to come down."

For material handling systems, the packaging of the card could be hardened for the industrial environment.

Smaller and lighter tags mean new methods of inventory control as they are stuck to the sides of incoming pallet loads or placed in open bins with inventory status on them.

As with other technologies, use of RF in the commercial sector for toll taking, parking access, commercial credit and other applications will get tags produced in high volumes. This will force the technology into a high production, high tech packaging approach that will lower the price.

We haven't yet heard from the Japanese on all this. They certainly have the same semi-conductor capabilities that we have and could become important in the RF/ID area.

RF/ID is a technology that can complement other code methods already in place. The only consideration today is the cost. There are plenty of RF/ID practitioners ready to develop systems and add value to RF/ID capability in making it produce the advantages it has promised. The only question is, at what cost and with what capabilities?

RF/ID: the official definition

To keep the radio frequency used for automatic identification purposes separate and distinguishable from that used for RF communication — terminals mounted on lift trucks, for example—the Automatic Identification Manufacturers, Inc., (AIM USA) offers the following official definitions.

Radio Frequency Identification (RF/ID) refers to systems that read data from RF tags that are present in an RF field projected from RF reading equipment. Data may be contained in one (1) or more bits for the purpose of providing identification and other information relevant to the object to which the tag is attached.

RF/DC: the official definition

Radio frequency Data Communications (RF DC) refers to systems that communicate data over radio link between a host computer and a data source such as keyboards, data terminals, readers for OCR, bar codes, mag stripes, RF/ID, etc. RF DC enhances the capabilities of hard wired communications without the restrictions of interconnecting wires.

There is a subcommittee within AIM that is expert on RF/ID. Chairman of this subcommittee is Ron Ames. AIM is located at 1326 Freeport Road, Pittsburgh, PA 15236

Industry Growth and Market Trends

The articles in this chapter focus on the development of the world RF/ID market as a whole, with some attention to the segmentation which already exists or is expected to occur as the market grows. Since Standard Industry Codes (SIC) have not been established for the Auto ID industry in general, nor for the RF/ID segment specifically, therefore no official market size or growth data exists. The following articles reference private studies which have attempted to derive these data from unofficial sources.

Comment: 2.1

The next article presents some of the data gathered and reported in the study "Radio Frequency I.D. : Technology, Markets and Applications", by Richard D. Bushnell and Ron Ames, published in February, 1988 by Cutter Information Corp.

As a summary of the book, the author fails to point out that the projected growth of RF/ID would only be possible if the industry provided sufficient capacity and support . This has not yet occurred.

2.1 RF/ID:

The right frequency for rapid automatic ID growth

Richard D. Bushnell, Jr.—Automatic I.D. News—© February 1988

What is it that ranges in size from a grain of rice to a pack of cigarettes, can be used to identify salmon in the wild or carry a complete manufacturing build specification through a plant or allow hands-free access to a secured area - and do all these covered with mud, cloth, or protective packaging?

It is RF/ID Technology (Radio Frequency Identification) which uses transponders—also called tags—essentially tiny radio receivers and transmitters that work in conjunction with a reading and transmitting device linked to a computer. The RF tag carries information ranging from one unique number, like a license plate, to thousands of alphanumeric characters. The tag can be read or have data written on it. If you want to draw an analogy to a bar code system, the RF/ID reader is like a bar code scanner and the tag is like the bar code label, but that is where the similarity ends.

Automatic identification needs go far beyond applications met with bar codes. This means that whole new vistas have been opened. In dollars and cents it means that RF/ID could—if production of the equipment and sales/engineering support is adequate—add as much as 50% to the anticipated $2.3 billion Auto. ID market forecast by Frost & Sullivan for 1990. This could make the Auto. ID market expand to $3.19 billion in 1990 or 1991—that is to say that RF/ID sales are estimated to be $58 million in 1987 and could be $1.34 billion in three to four years.

How could such a giant grow from such a tiny acorn? Several reasons:

1. The concept of automatic ID is now widely accepted and is broadening every day.

2. The computer power to use information gained from Auto. ID is getting cheaper.

3. RF/ID costs will continue to decline.

4. Huge applications are surfacing which are distinctly different from bar codes.

Table 2-1 shows the growth of all forms of Auto. ID from 1986 to 1991. The dollar volumes show the market size by application group: industrial, government, service and retail. To that we added the potential RF/ID sales. RF/ID is only .5% of total sales in 1987 but could be closer to 28% in 1991.

Table 2-1   TOTAL AUTO. ID MARKET
(millions of dollars)

All Other Forms	'86	'87	'88	'89	'90	'91
Industrial	290.93	395.66	542.06	731.78	1017.18	1393.53
Government	169.26	223.42	297.15	371.44	508.87	686.98
Service	69.42	92.33	124.64	169.52	230.54	308.92
Retail	252.64	310.75	391.54	516.83	713.23	984.26
Sub Totals	782.25	1022.16	1355.40	1789.57	2469.82	3373.70
RF/ID Hi and Low Freq.	N/A	58.23	158.05	410.87	720.40	1340.75
Totals	**782.25**	**1080.39**	**1513.45**	**2200.44**	**3190.22**	**4714.45**

All Other Traditional	Table 2-2 IMPACT OF NEW APPLICATIONS (millions of dollars)				
	'87	'88	'89	'90	'91
Sales	52.33	78.10	107.01	173.90	210.00
New Applications					
Auto Lisc.					
Auto VIN					
Baggage					
Laundry					
Medical					
Tires					
Tooling					
Sub Totals	5.90	77.95	303.87	546.87	1130.75
TOTALS	**58.23**	**156.05**	**410.87**	**720.40**	**1340.75**

Characteristics of the RF/ID technology

RF/ID technology is currently employed in a variety of applications where target orientation or line of sight between reading element and encoding element are random.

Another unique feature of the RF/ID technology is its ability to have information programmed or reprogrammed into it without physically touching the encoded element. This allows for very broad and new applications.

Aside from the traditional use of RF/ID, there is potential in material movement in industrial applications, security, employee identification and animal identification. Information we have obtained shows extreme interest in completely new applications which include automobile licenses, automobile vehicle ID numbers, baggage, cargo, medical, tires, tools and transportation—see Table 2-2.

Many of the Auto. ID technologies have overlapping applications. For example, a bar code label could be used to identify an automobile as it passes a laser-based bar code reader. So could an RF tag. On the surface, it would appear that the two technologies compete. However, if we're to look further into the application, we see that there are times that one would have distinctly different benefits. For example, if we wanted to automatically identify the automobile during the manufacturing process, including assembly and painting, we might find bar code to be inappropriate, since it could be obliterated by the manufacturing or painting process. Research shows that the different technologies do not compete for the same dollars.

RF/ID systems offer the following general benefits which are quite different from other Auto.ID methods:

- RF systems are not limited to line-of-sight like bar codes. This means that the RF tag can be placed under, in, or be covered by the container or product on which it is being carried. Furthermore, dirt or other opaque substances will not affect the readability of the tag.

- RF tags can provide a read/write capability, which means the user can change information related to the status, contents or physical condition of the item carrying the tag.

- RF tags can carry large amounts of data. There really is not a good way to compare this to bar codes, since data density in bar codes is an optical matter relating to characters per inch. Data bits in the RF tag are electronic - not a linear dimension relationship. An RF tag carries from 8 binary bits to thousands of 8-bit bytes of data.

- The Operating range of RF systems can be far greater than that of optical systems. In some cargo handling applications, RF systems operate at distances of 100 to 150 feet with a battery in the tag, or 20 to 30 feet without.

- RF tags are difficult to counterfeit. Since bar codes use an optical system and are designed to be easy to print, they obviously can be duplicated. There are some techniques to secure bar code symbols against counterfeiting; however, the nature of RF tags make them significantly more difficult to replicate.

The striking difference between the bar code companies and the RF/ID companies is that the RF/ID companies derive significant income from the tag or transponder business, that is, the item being read. Typically, bar code companies derive little income from the labels being read.

RF/ID tags are really Integrated Circuit (IC) chips which are available only from companies who possess advance manufacturing technologies. Some of the technology will be protected by licenses and patents. This manufacturing exclusivity provides a real opportunity for market control. Some industry experts believe there is an opportunity for a chip manufacturer to supply tags to a variety of equipment supply companies who would incorporate that mass-produced, low-cost chip into their product. If this does not happen, then competition will be based on market share by application niche or vertical market.

Pricing

The following system and tag values are based on revenue to the manufacturer. They do not include supplier mark-up or added sales commission. First, the fundamentals will be explained, then the range prices will be shown.

To determine sales revenues, cost models have been developed to reflect the types of systems employed. The types differed in average cost of tag and/or reading and interface equipment along with the ratio of readers to interfaces and the ratio of tags to reading points.

To provide an idea of the equipment costs, here are two examples of service models developed. The first system employs 45,000 tags. The average value was $27 and the total value for those tags was $1,215,000. The reader (RDR) and interface (I/F) value is $4,575,000. That value is based on 750 I/F at an average value of $2,500 which were linked to 3000 readers with an average value of $900. This was based on a ratio of 4 RDR to every I/F, and 15 tags for every reading point.

Another example employs 1,750,000 tags with an average value of $7. The total value is $12,250,000. The RDR, I/F value is $17,500,000 based on 35,000 decoder I/Fs with an average value of $500. This represents a ratio of 1 RDR to every I/F, and 50 tags for every reading point.

Another way to look at equipment cost is by technology type. The models used are broken into low and high operating frequency. With low frequency, the value of the tags ranges from a low of $7 up to $160. The difference is largely in the packaging. Virtually all the manufacturers said the price is coming down. The equipment value range is also quite varied, with $70% of this type having a value of $500 and the maximum cost being $1,800.

With the high frequency equipment, the greatest number of tags is in the $35 range with a high of $150. Scanner system costs range from $1,200 to $5,600 because of the high cost of the reader points as well as the RDR I/F. Keep in mind that the High Frequency Systems provide significantly different performance characteristics than Low Frequency and that justifies the increased cost.

Conclusion

Keep in mind the following points when considering an RF/ID system:

1. The potential market is approximately four times lager than the current capability of the RF industry to support it.

2. Potential applications are broad: local, state and federal governments—including defense, commercial and industrial fields.

3. Markets vary from niche areas like commercial laundries to grand scale markets like automobile vehicle identification numbers.

4. Solid state technology lends itself to high tech chip manufacturing.

5. An enormous value-added market exists for software and systems to exploit the information that the technology can provide. This automatic ID information is in a form never available before.

Productivity improvements and industry experience lead one to believe that RF/ID offers a large opportunity in a soon-to-be billion dollar industry. The only question is which companies and which applications will profit. The answer lies with you, the practitioner in this industry.

The author, Richard D. Bushnell, Jr., is president of Bushnell Consulting Group and has been heavily involved in Auto. ID since 1969. His experience includes directing marketing efforts for several major Auto. ID companies, and developing corporate strategy and system design. He now serves as Administrator of the Industry Bar Code Alliance (IBCA) which represents over 30,000 manufacturer, wholesaler and supplier locations dealing in non-retail items. Bushnell has authored "Getting Started with Bar Codes: A Systematic Guide" and contributed editorially to several trade publications. Automatic I.D. News, February 1988

Comment: 2.2

Although the author of the following article does not quote the specific RF/ID figures from the AIM Europe Market Research Report mentioned in the article, the growth rates are reported to be roughly proportional to those in the preceding article. He does develop the character of the European market and the expectations which must be met in order to successfully participate in it.

2.2 GROWING STRONG

A EUROPEAN PERSPECTIVE ON RF/ID

Stuart Evans—ID Systems—© December, 1987

An increasing number of people in the world of automatic identification are beginning to share some of the excitement about the potential scale and impact of RF coded tag systems. Unlike bar code technology, RF coded tags are at a much earlier stage of industry development. This most vividly manifests itself in a lack of standards: RF tags and readers must still be purchased from the same manufacturers—tags from one manufacturer cannot be read by another manufacturer's readers. For this reason, choosing an RF coded tag system or vendor becomes particularly important. And, since many RF coded tag systems come from Europe, understanding the European market and how it differs from the U.S. market is essential.

The European market is strong and growing rapidly. The recent AIM Europe Market Research Report estimates the market will reach $253 million by 1992, growing from $50 million in 1987 to $75 million in 1988, $110 million in 1989, $153 million by 1990, and $200 million by 1992—these figures from AIM Europe, despite average price drops of 30 percent. This growth will come from solid, steady expansion of the market as a whole rather than any single new application. If any major new applications take off, growth could be even higher.

Most local demand is supplied by European-based companies. Despite the slow evolution of the EEC towards a real "Common Market," Europe remains primarily a group of national markets. Not surprisingly, Germany - site of Scan Tech Europe '87 in Dusseldorf - is the largest national market, with 23.3% of the European market, with the UK (20.7%) and France (13.6%) not far behind. Each country has its own language, culture, and approach to business. It is dangerous to assume that English—or American!— is the "lingua franca" that it is in, say, computers or banking. Product literature—both sales and technical—needs to be available in several languages.

In addition, each country has its own unique set of rules and regulations with which any RF coded tag system must comply. These fall into four major categories: PTT approval to use a particular part of the radio spectrum at particular power levels, radio frequency interference (RFI) approvals, electrical safety approvals, and environmental approvals. Because this is such a new market, it is often a major task to identify who the relevant approval bodies are in each country and which regulations are relevant. Testing laboratories vary from country to country, and because the products are so new, equipment that passes in one country may fail the same tests in another. Gaining a comprehensive set of product approvals, then, takes patience, persistence, and in-depth technical resources.

An added complication is that the relative importance of acquiring approval in each area varies from country to country. In Sweden, for example, Semco (electrical safety) approval is generally regarded as vital. In contrast, launching a tag system in Germany without PTT approvals is essentially impossible.

The European market is further characterized by the presence of large companies—Philips, Bosch, Brown-Boveri, GEC, and Siemens all offer systems. In contrast, the only large companies in the tag business in the U.S. are Allen-Bradley and Ingersoll Rand—via its Schlage Electronics subsidiary. Most of these large European companies have concentrated on industrial/factory automation applications and have probably been more conservative in the projects they have pursued.

By avoiding "mega systems" projects, European companies have also avoided the associated pitfalls:

- Projects get too big to handle; total project cost can easily be ten times the cost of the tag subsystem.
- Getting a slow start is impossible. The 80/20 rule instead of the 20/80 rule applies: you have to make 80 percent of the investment before you get even 20 percent return.
- Benefit/investment mismatch: a large system makes economic sense overall, but requires a group to buy tags even though they do not benefit directly. Altruism apart, this only works via some kind of transfer pricing mechanism or sanction power.

So, by not becoming entangled in these pitfalls, any combination of which can stop the show, European companies have gained solid experience, learned the strengths and weaknesses of their system, and built a steady business base.

Another intriguing aspect of the European market is the possible convergence between RF coded tags and smart cards. One of the most successful European tag systems is the French Statec system for industrial automation, which operates at a short range of a few centimeters. Tag systems that are now coming out for tool identification are also short range. France's commitment to all forms of "Informatique" has greatly stimulated the market for IC/smart cards.

Because of user concern about the contacts for reading and writing to these cards, some companies—like GEC—have introduced non-contact smart cards that operate at short rangers of 20 mm. On the margin, a non-contact smart card is hard to distinguish from a short-range RF coded tag. How far and how fast these convergence will go remains a question.

The European segment appears to differ from the U.S. market in the area of access control. In the U.S., Schlage pioneered proximity access control, offering a tag with relatively short range - typically less than 10 cm. Many new U.S. tag companies aiming at this segment offer a comparable range. In contrast, European companies offer much greater ranges—1 meter or more. This longer range permits hands-free operation, making automatic vehicle identification a practical possibility. All this adds up to greater user convenience. As one observer put it, "In the Old Days, we had card-based access control. Then short-range proximity became available. This was like moving from radio broadcasting to black and white TV. Now we have systems that provide real hands-free operation. This is like moving from black and white TV to color TV. Once you've got used to real hands-free access control, you will never go back to the old system."

What do all these differences amount to? The RF coded tag systems market in Europe is thriving, offering an incredibly diverse range of applications. So when choosing a system, be sure to review the options available to you from Europe.

	3

Technology

The articles in this chapter cover the technologies used in the complete range, of what is considered to be RF/ID, in greater depth than the previous articles. They will not make you an expert but they should stimulate you to ask the questions most relevant to your needs. As you read, you will become aware that several classes of technology, not just **semiconductor chip technology**; but **materials and process technology**,—including the deposition of metals,use of ceramics, composition and use of plastics; **communications technology**— including local area networks, wireless modems and RF data communication terminals; **power generation and storage technology**—including batteries and solar cells, on-chip power conversion and regulation and other technologies which can contribute significant leverage to this industry.

3.1 RF/ID Systems

Ron Ames—ID Journal—© July/August, 1988

Let me first present the technology upon which all current radio frequency identification products are based. From this vantage point we can examine considerations relevant to the application of RF/ID to real-world problems. We can then identify the applications best suited to RF/ID technology, and review new technology which will have a significant impact upon this industry.

The Technology

We begin by looking at the electromagnetic principles that makes radio frequency identification (RF/ID) possible. The basic principles were primarily discovered by Michael Faraday, Nikola Tesla, and Heinrich R. Hertz prior to 1900. All three were working to discover the fundamental behavioral characteristics of electrons, the properties of materials which function as conductors and insulators, and the interrelationship of electricity and magnetism.

From them we know that when a group of electrons or current flows through a conductor, a magnetic field is formed surrounding the conductor. The field's strength diminishes as the distance from the wire increases. We also know that when there is relative motion between a conductor and a magnetic field a current is induced in that conductor. These two basic phenomena are used in all low frequency RF/ID systems on the market today.

When the conductors are formed into a coil, the intensity of the field is enhanced, by keeping the field closer together, and through the phenomena of self inductance, in which the lines of force generated by the conductor intersect the same conductor at multiple points as the field builds and collapses. If the current varies through the coil, a magnetic field is developed around the coil which varies in concert with it. The building and collapsing of the field induces additional current to flow in the conductor which is called self inductance.

The next interesting phenomena occurs when a second coil is placed within the field generated by the first coil. A current is also induced in the second coil, which corresponds to the current in the first coil, that is, they increase and decrease together. This phenomena is called mutual inductance. This is the principle upon which all transformers are based. Many of them in common usage have a ferrite or iron core passing through the centers and even surrounding both coils. The reason is that the iron or ferrite materials serve to direct and concentrate the magnetic fields such that they have a greater affect on the coils.

If one of these coils is to be located in a tag and the other one is a portion of a reader, you can see that it is impractical to tie them together physically using a common iron core. For that reason, we must tolerate a lower efficiency of coupling between the two coils in RF/ID applications.

Those who are familiar with radio communication will immediately see that this is different from normal radio transmission. We are using what is called the "near field effect" rather than "plane wave transmission."

They may also recognize that this form of coupling decreases in efficiency more rapidly with distance than does plane wave transmission. This is mathematically described as decreasing as the inverse of the distance cubed, rather than as the inverse of the distance squared, as is the case in plane wave transmission. This shorter range can give us the advantage of greater precision if we are trying to determine the location of the tag, in addition to its identity.

Some RF/ID systems do use the plane wave for communication but since most of you are familiar with radio and TV, I will not rehash the subject.

Terminology/System Elements

Let me now develop some of the terminology used in radio frequency identification. The coil that transmits to the tag and receives the signal back from the tag, when combined with the electronics to do both functions, is called a scanner or reader coil. The electronic portion is sometimes called the transmitter (or exciter), the receiver, or the transceiver. When all of this is combined with digital electronics, usually including a microprocessor, it is called a reader. In Europe, it is also called an interrogator.

Typically, the tag also includes a coil, transceiver electronics, control logic, and some form of nonvolatile storage or memory. In most cases, the reader or scanner broadcasts a frequency to the tag, which serves, at a minimum, to notify the tag that a reader is present, and wants to receive its identity or perform some transaction. The tag usually responds with coded data modulated on a different frequency. The reader receives this signal, extracts the data from it, and presents it in visual form to a human, or electronically to a host computer.

The methods of encoding information, or modulating it on the carriers to and from the tag, include all of those used in other forms of radio frequency communication. The carrier frequency sent to the tag may also have two additional purposes:

1) it may provide the digital clock for the logic in the tag and
2) it may provide the energy to power the circuitry in the tag.

At this point an obvious question would be, 'what happens when more than one tag is within normal operating range of the scanner?'

In most of the products currently on the market, the scanner would simultaneously receive return signals from all of the tags within range and would not be able to differentiate between them until one or the other moved just out of range.

In some cases, if the tags came into range, or left in a sequential order, they might be successfully read. In a few cases, a delay between periodic responses from the tag can be programmed into the tag, providing a statistical probability that, at certain times, only one tag would be transmitting, and could therefore, be clearly understood. There is an obvious practical upper limit to the number of tags which could be distinguished in this way, or a practical minimum time to read all possible tags which were within range. Another approach to solving this problem is to design tags that only respond when they are polled, or receive their unique identification from the reader. Very few low frequency systems use this approach, because their range is typically rather short. For this reason, low frequency systems are not currently categorized as polled or unpolled.

Active/Passive

RF/ID products are categorized by their source of power. Active tags are those which contain a battery. Those that contain batteries may use them in different ways. For example, some tags use the battery to power all of the elements, including the transceiver, the logic, and the memory. When batteries power the memory, it is usually made up of CMOS static RAM which uses very little power, except when being read or written. Some tags use the battery only to power the memory, and convert part of the power contained in the carrier from the reader to DC to power the logic and transceiver.

Still others use this converted power for everything including the transceiver, the logic, and to write into a non-volatile memory such as EEPROM. These are called passive tags. Other forms of non-volatile memory may be used such as EPROM or laser-scribed, on-chip links or in the near future a new technology called ferroelectric memory may be used.

The typical range for low frequency active tags is 3-15 feet or 100-150 feet for high frequency tags. The typical range for passive tags is less than 18 inches for low frequency and 15-30 feet for high frequency tags. Longer ranges are achievable in both cases by using larger or more efficient antennas.

Programming

Another way of categorizing systems is by the method of programming information into the tags.

When a tag can only be programmed at the time of manufacture, it is called factory programming, and results in a read only tag.

In cases where the tag can be programmed by plugging it into, or placing it very near a programmer, it is called field programming. This usually requires the tag to be removed from the object it identifies. When information can be written into the tag in its normal operating location, it is called in-use programming of a read/ write tag.

Capacity

Level I

Tags can also be classified by capacity or the number of bits stored in the tag. The simplest tags—Level I—are those which are typically used in retail stores to prevent shoplifting and in effect contain one bit, indicating the presence of a tag.

Level II

The next level of capacity—Level II—are tags used for identification of people or objects. They range in capacity from 8 bits to 128 bits and are usually either factory or field programmable.

Level III

The third level—Level III—is called transaction/routing tags that include from 48 bits to 512 bits of information, in addition to identity. This information is usually status, flag, or location information in an encoded form where each bit has a unique meaning.

Level IV

The fourth level—Level IV—is called portable data bases. These usually carry from 256 bits to 256 thousand bytes of information in text using ASCII alphanumeric code.

Intelligence

An additional method of classification is just becoming necessary. It relates to whether the tag has information processing or decision making capability. They are, therefore, classified as intelligent or dumb. Intelligent tags are implemented by incorporating a microprocessor in them. This class, of course, includes radio frequency versions of Smart Cards.

Orientation

The most favorable orientation in an inductive system is when both coils have the same axes, if they are similar in size, or tangential, if they are dissimilar in size. The least favorable orientation is when their axes are perpendicular. The least favorable attitude may result in a complete inability to read the tag, but more commonly, the read distance may be decreased from optimal by 30-50 percent.

The separation between unpolled tags is recommended to be twice the normal reading distance to avoid confusion.

Frequencies Used

The scanner carrier frequency ranges which are classified as low frequency are 500 kHz and below. There are a few at 400 kHz, but most are below 150 kHz due to regulations in Europe and the United Kingdom. The AM radio band in the U.S. begins at 560 kHz, going up to 1650 kHz and most companies have avoided conflict with it.

High frequency systems typically use 908-920 MHz, 1812-1830 MHz, and 2.45 GHz, but may be even higher.

Applications

Now that we have covered the technology and how products work, we should look at the potential applications to see which are most suitable for RF/ID products.

The industry has attempted to characterize applications in three ways. The first of these is by the objects that are identified. This is the most common method, and we will use it. Two other methods are, abstractions that look at classes of characteristics of the applications, and by the function being performed.

Objects

Objects are broken up into things, people, animals, vehicles, places and money.

There is an obvious redundancy between things, vehicles, and money. The last two are also things, but they are very important to us and have several unique requirements deserving special attention.

Things

Most of the things that we would desire to identify, or attach information to, are fabricated items in the broadest sense of the term. This would include manufactured goods and even art objects. Since most of the items in this category can be approached, or have a reader located nearby, they are good candidates for low frequency RF/ID systems. Exceptions to this would include cargo containers, or other items requiring a longer read range.

Since the carrier frequency provides an upper limit to the rate at which data can be transferred, low frequency systems have an inherently lower data rate. Objects that are moving rapidly or require a large amount of data to be transferred, may be better candidates for high frequency RF/ID systems.

People

The identification of people usually falls into two categories. One is physical access to a facility, and the other is the right or privilege to use or own a resource. Physical access to a facility based upon an RF/ID badge or credential implies an 18 inch range as a requirement, if the credential is to be read anywhere on the body, or in a purse or briefcase carried by the person. This is near the upper limit of passive low frequency systems, but is well within active systems. The use of resources usually implies a shorter range requirement and is, therefore, well matched with passive low frequency systems.

Animals

The identification of animals should not be confused with the requirement to locate animals feral in the wild or domestic animals on the range. RF/ID assumes that the animal is confined and approachable and, therefore, requires a similar operating range to people. An example would be identifying cows in a milking parlor. Animal identification is almost exclusively a low frequency application.

Vehicles

Some vehicles represent a freedom of movement that may be beyond the capabilities of a low frequency system, unless the tag is read from an antenna buried in the roadway, or suspended above vehicles of known or limited height. They may also be read from the side, if islands or curbing limits the required reading distance. Most of the currently installed vehicle identification applications use high frequency systems.

Places

The identification of places may be as part of an automatic guided vehicle system in a factory, an automatic station announcer on a light rail system, or key locations or assets that a security guard must check on his rounds. Most of the systems using RF identification use low frequency systems, however, most of the systems currently installed for this application use older technologies.

Money

Radio frequency smart cards have only recently been introduced and the current predominating technology in money or transaction applications is magnetic stripe technology. This certainly implies that even short range RF/ID is a major step up in convenience from what is currently being used. The ability to use imbedded intelligence to handle multiple accounts will free users from having to have large numbers of cards. Significant advantages in card life and security will help these products penetrate the market.

In summary, low frequency technology is suitable for more of the potential applications, as well as, those that represent the highest potential sales volume. As a result, there are significantly more companies and products based on low frequency technology.

New Technology

One new technology which I mentioned earlier is the RF smart card. At least three companies, AT&T, GEC, and Electo-Galil have demonstrated smart cards, or devices with smart card functionality which are powered and communicate via signals at radio frequencies. Several others have announced their intention to provide complete products—e.g. Arimura, Valvo, etc.—or chips—e.g. Dallas Semiconductor, Single Chip Systems, Catalyst, etc.. Another new technology is flash EEPROMS. This technology allows all data to be rewritten a few times, ten or so. Since ALL of the data must be rewritten in order to change any of it, one approach is to invalidate the old data and add the new. This was the approach used back in the days when punched paper tape was used as a storage media. If the media is cheap enough, there is no better alternative, or if the data is very rarely changed, it may serve.

A fourth technology is ferroelectric memory, also mentioned previously. This offers nonvolatile storage which can be rewritten orders of magnitude faster than EEPROMs and more times than EEPROM. Even more important, in this industry where minimizing the voltage and power requirements of the tag are crucial in meeting the application needs and the competition, it uses lower programming voltage and operating power. It is also possible to fabricate large on-chip capacitors, and since the storage element stacks on top of a transistor, smaller cell sizes are not possible in any other technology currently available.

Although not currently available, significant progress is being made in materials technology for on-chip interconnect. Cheap metalization, more layers and perhaps someday, superconductivity at normal operating temperatures, will all be very important to this industry.

The use of on-chip coils for powering and communicating with a Smart Card chip is now patented and being implemented. This is undoubtedly the approach requiring the least packaging, i.e., zero pinouts.

With such advances in technology such as these becoming available as chip level products to anyone who has a need or a business plan, the increased understanding of appropriate applications, and the ability to plug RF/ID products into existing system interconnectivity and software (which have also been making great strides), the RF/ID industry seems to be well positioned for rapid growth.

Comment: 3.2

The following article accurately describes the current state of smart card development and forecasts the direction of future development. One error concerning the range at which the NEDAP mini-label can be read crept in. It should read "It can be read at a distance of 2 inches when not in the presence of metal......".This tag is designed to imbedded flush with the surface of a steel machine tool adapter and therefore achieves a much shorter range than that described when used in it's intended application.

3.2 SMART CARDS:

WALLET-SIZE SOLUTIONS

Stephen Seidman—Automatic ID News—© October, 1987

A variety of applications are emerging, requiring various standards for smart cards.

This year marks the fifth anniversary of the first live trials of smart cards, which were conducted in three small cities in France. Since that time, during this relatively short period, there have been dramatic changes in the technology, in the applications, and in the packaging. And we can already see more changes just over the horizon.

Up until two or three years ago, there was a single-minded worldwide effort to develop a smart card standard that all nations would agree to use. Through the International Standards Organization (ISO) these efforts are continuing. However, it no longer represents the single-mindedness of earlier years.

This evolving standard is being seen, more and more, as something that the financial community needs for transaction processing applications, but which, perhaps, many other applications do not require. As the ISO standard comes to be seen less and less as a universal requirement for smart card design, the extensive deviations from the standard design make it quite difficult to any longer answer with ease, the question, "What is a smart card?"

What is a smart card?

Five years ago, a smart card was a credit card-sized piece of plastic with one, or perhaps two chips embedded in it. There was an eight-bit microprocessor—or functional equivalent—and either an 8 K-bit or 16 K-bit read-only memory—Erasable Programmable Read Only

Memory—EPROM. The memory could be written into only once, but could be ready any number of times, as required.

For financial applications, this was wonderful. Transactions could be recorded and could not be altered, and a permanent audit trail was available for review at any time. The main drawback of an EPROM memory is that when it is full, there can be no further transactions, and so the smart card must be discarded.

There are many people developing applications for which an EPROM memory, which eventually leads to discarding the smart cards, is unacceptable. If for no other reason, it's an expensive procedure. They would like to have the standard card designed around an EEPROM— Electrical Erasable Programmable Read Only Memory—chip. This is unacceptable to ISO because of legitimate concern that a rewritable memory, such as an EEPROM, could have its contents fraudulently altered.

Further, they argue, EEPROMs cost enough more than EPROMs that it may well be cheaper in the long run, to throw away the card. And, finally, until just recently, no manufacturer had been able to supply an eight-bit microprocessor-plus-EEPROM for the smart card on a single chip. And everyone knows that if it isn't a single chip, the wires between the chips can be probed and read, and the contents of the memory altered. Stalemate.

Multiple designs.

It is now obvious that there will not be a single description or standard for smart cards, and that there will be a multiplicity of designs, for a multiplicity of applications. Where appropriate, it will obviously be cost-beneficial to hang on to the coattails of the financial giants, Visa and MasterCard.

Together, they will probably be issuing hundreds of millions of smart cards per year, and possibly as many as a million terminals, within five years. For those who choose not to wait, or not to use the standard, the range or options regarding card size, alternative package designs, numbers of chips, memory type and size, physical interfacing— contacts or not—and level of integration, is increasing rapidly.

Card Size

The requirement for smart cards to have the same physical dimensions as a standard plastic credit card grows out of the fact that the driving application for the development of smart cards has been the replacement of standard magnetic striped credit cards. Considering the potential for replacing about a billion bank credit cards now in circulation around the world, and comparable numbers of debit (ATM) cards and retail store and gasoline cards, the power of the application to drive the market is obvious.

Nevertheless, smart cards are already available which conform to length and width specifications of the ISO standard, but which are thicker. Usually, they are only two to five times thicker, ranging from 2mm to 4mm, instead of the standard .79mm. While these would be awkward to carry in a wallet, they are not intended for that purpose. Generally speaking, they are cards with large memory capacities, or large numbers of special function chips, or cards which are contactless and sturdy, in order to function in potentially corrosive factory environments.

The Nippon LSI (Japan) card, for example, is a non-contact card having 32 kilobytes of memory, housed in a card which is 3.8mm thick. This card is used by Hitachi Seiki to program its numerically controlled machines on the factory floor.

When is a card not a card?

For several years, Datakey, Inc. has been supplying smart card-type chips with smart card-type functions in packages that look like plastic keys, and in packages which look like—and are used experimentally as—military dog tags. The keys have also been used to implement "smart" kidney dialysis machines, and "smart" fleet maintenance and refueling operations. And, while nobody would call them smart cards, they always get lumped together in discussions of where smart card product development is headed.

More recently, Mars Electronics announced a Smart Coin, plus a line of tags and cards. They are all being designed to work with a non-contact interface. Mars, which is very active in vending machine coin collection equipment, designed the coin so that it would fit into the standard slots and channels in coin telephones and coin-operated vending machines. As the coin, which as a prepaid value, drops through the channel, it is debited by the amount of the phone call or the vended item, and returned to the owner. When the coin's value is used up, it can either be discarded or recharged and reused. Using the coin would eliminate the requirement to replace the coin mechanisms with smart card handling mechanisms when that evolutionary step in vending machines occurs in the U.S. It is already beginning in Japan.

The smart card connection.

As originally designed, smart cards have eight closely spaced gold-plated contacts flush with the front or rear surface of the plastic. Six of these contacts are currently functional, and two are spares. When such a smart card is placed into a reader/writer, all six functional contacts must make a solid electrical connection. Dirt, grease, oxidation, wear, coffee, coke, and who knows what else are potential inhibitors of those contacts being effectively made.

Several companies around the world have developed non-contact smart cards to solve this problem. AT&T, in the U.S., has developed a standard-sized telephone smart card which magnetically couples power into itself from AT&T's modified Public Phone Plus and capacitively couples data in two directions between itself and the telephone. LSI Nippon is the key non-contact vendor in Japan. In England, GEC has developed a card in standard ISO dimensions, and it is being tried in a pilot program with Midland Bank. On a totally different track, a Dutch company, NEDAP, has developed a non-contact minilabel, not much larger than the size of a large aspirin tablet. It contains an EEPROM memory. It can be read at a distance of 28 inches and can be written into at just under .2 inches. It is to be introduced at the SCANTECH '87 Show.

All of these non-contact arrangements also solve another problem of some considerable concern, which is the potential damage to be done to the smart-card-with-contacts by the buildup of static electricity at the contacts. A good shuffling walk across a carpet on a dry warm day could create a static buildup of tens of thousands of volts.

The ultimate card is its own terminal.

Before you decide that this one is too far out, think of the Japanese pocket calculators you have seen lately, selling for less than ten dollars. They contain, in addition to their integrated circuits, integrated keyboards, displays, and one-to-five-year batteries. Yes, they are thicker than the standard plastic credit card, but only temporarily. The proposed Visa Super smart card, which is to be delivered by Toshiba (Japan) next year, is expected to cost the individual banks ten dollars less. It is designed to meet ISO standard dimensions, including thickness. In use, it generates its own credit authorization number, based on the status of the user's credit, the amount spent since the last payment, and the amount of the proposed purchase.

It can be programmed to detect deviations from the owner's usual spending habits, and to alert a merchant to perform a mandatory on-line check. If the card has been reported stolen, or is being used in a fraudulent manner, it is automatically disabled by the authorization system. Otherwise the card need not be connected to a terminal more than once a month or once every-so-many transactions for a routine check and data synchronization. A similar smart card, manufactured by Oki Electric (Japan), is being tested in Japan by Fuji Bank. It used a solar system for power instead of a battery.

To speak of the future of smart cards without being able to say what a smart card is, must surely be reaching. So, we will reach. The standard financial smart card will, within five years, become a commodity item, manufactured offshore. Within five years after that, manufacturing will reach levels of several hundred million per year.

Meanwhile, a variety of other applications will emerge, requiring different sizes and shapes of smart cards. Applications will include medical—perhaps everyone will carry a card with emergency medical information in it and automotive—perhaps every car will have a smart card to open the car door, turn on the ignition, keep track of driving records, maintenance required and completed, electronic maps, and vehicle registration. Individuals will have "smart" driver's licenses— which will fit neatly into those readers in the highway patrol and police cars, and, of course, security applications—to get into home, office, gym lockers and safety deposit boxes.

There will be prepaid cards or coins for use in public transportation, telephones, vending machines and parking meters. Smart card subscriptions to theater, sporting and entertainment events will operate turnstiles at the entrances.

Smart cards will get larger and smaller, have more and less memory, both erasable and non-erasable, and eventually will come in as many different sizes, shapes, capabilities and prices as any other computer.

Comment: 3.3

The following paper addresses tags which have a large data storage capacity and relates them to factory applications. There are also many applications for this class of tags outside of the factory. Some of both of these types of applications are discussed in later chapters.

The author covers much of the same ground as the first article in this section, which is not surprising since much of the basic terminology was developed in the context of AIM.

Some will challenge equaiting data with intelligence, however their is no valid challenge to having read/write data in a tag.

He goes on to provide additionel detail about operational considerations of these large capacity tags, which may also be called portable data bases or communicationg electronic documents.

3.3 TRANSACTION AND PORTABLE DATA BASE

RADIO FREQUENCY SYSTEMS

David Wm. Draxler—ID Expo—© April, 1988

How can you improve manufacturing and distribution efficiency while reducing information handling and maintain maximum flexibility and quality?

We would like to explore this question by looking into innovative Radio Frequency Information Equipment solutions.

HISTORY OF MANUFACTURING CONCEPTS

Way back at the dawn of manufacturing, family businesses or "craft" shops prevailed. These craft shops were small businesses which generally produced a single product or type of product.

To expand production meant that more employees had to be "hired" into the family. Production power was generated by either human strength or some mechanical device like a water wheel.

Although high product quality and flexible design characterized these businesses, problems such as low productivity, lack of standardization and scheduling problems were prevalent.

Enter the industrial revolution and the advent of the factory. Now we were able to harness the water wheel and attach it to a central line shaft to operate many machines in a production line.

With high productivity manufacturing assembly lines initiated by Henry Ford in the beginning of the 20th century, medium and large sized companies now were able to mass produce items with a significant degree of standardization. High production rates, however, brought with them a loss of quality, pride-of-workmanship and limited assembly options. Work-in-process was afforded greater ease in scheduling but created higher inventory levels.

Coupled with the mass-produced assembly lines we still had smaller job shops for support which eased the problem of flexibility and quality but brought with it low production and much high cost.

COMBINING CRAFT SHOP

THE FACTORY CONCEPT

The question was raised,"How can we combine the good qualities of job shops with the mass production capabilities of the factories?"

Enter the Data Processing Department. We now are able to improve productivity by bettering work-in-process scheduling, reducing lead and material handling time while lower inventory level requirements.

New challenges now arise. It becomes costly to identify work-in-process, we have difficulty in communicating between data handling devices and we begin to generate volumes of paper. In addition, now we are faced with the possibility of system shut-down due to equipment failure coupled with loss of manufacturing and management data.

THE RIGHT THING-in-THE RIGHT PLACE-at-THE RIGHT TIME

High productivity and assembly obviously requires work-in-process and storage facilities and we must efficiently use both. By increasing the manufacturing and assembly-to-storage ratio we could save millions of dollars. The main thing is to get the right thing, in the right place, at the right time. The ability to do this requires that each product is identified and proceeded along the line according to established procedures.

The manufacturing industry has previously accomplished this by attaching a manual paper work order on the product or load carrier. From there it is hoped that the work order remained with the product throughout manufacturing and storage—usually it does.

Advanced technology then introduced bar codes to provide the product with a "license plate", an identification tag which a computer can read for a machine or operator—the degree of automation advances! Awkward and difficult routine work tasks are left to machines and operators are given more advanced, skilled tasks. Industrial efficiency increases and profitability goes up.

Bar codes do, however, have limitations. They can only accommodate a small quantity of information; data can not be changed; dependency on a master-controller computer increases; and they are sensitive to dirt and heavy handling.

SOLUTION

TRANSACTION
AND
PORTABLE DATA BASE
SYSTEMS

Radio Frequency (RF) based Transaction and Portable Data Base Systems provide a new alternative to manufacturing information problems. Small, durable, flexible data modules with sizable memory capacity provide efficient mini read/write data files for each product. By lifting the file from the central processing unit, we now have the capability for each product to be intelligent and carry its own data while being manufactured or assembled.

Communication between devices, such as machining centers, now becomes easier because they are reading and writing common data in a common format.

This technique has proved to be an effective tool in automation around the world. They have provided a wide range or applications and offer an almost unlimited scope of corrective solutions for controlling flow of materials through manufacturing processes.

COMBINED ADVANTAGES

We now gain the advantages of the above individual systems and reduce the disadvantages.

Portable Data Bases are able to carry—order systems such as:

> Bill-of-Materials
> Assembly Instructions
> Product Status
> Work-In-Process Status
> Distribution Information
> Data Processing Information
> Scheduling Information
> Management Information

In addition, we have the advantages of the combined assembly systems:

> Flexible work flow
> Flexible automation
> High Productivity - High Quality
> Shortened Lead Times
> Less Materials Handling

Data processing advantages afforded by Portable Data Base Systems include:

> Improved/Easy Scheduling
> Improved Inventory Levels
> Current Management Information
> Current Work-in-Process Information

One disadvantage, or better stated, challenge, to working with Portable Data Base Systems is we must rethink our concept of manufacturing and learn new processes. This however will give us the opportunity to improve plant design, manufacturing flow and data process design.

PORTABLE DATA BASE BASICS

Generally, an RF portable, programmable data base system is comprised of memory modules and read/write processors.

As stated before, we can now decentralize the data files—intelligence—to the product or load carrier. The memory module now provides the capability of updating or completely changing information and instructions during transportation via the radio frequency read/write units either as stand alone units or in conjunction with a central processor.

Each product carrying its own data now communicates directly with the work station, worker or controlling systems in the manufacturing and material handling process.

Portable data base memory modules are available in several physical sizes and memory capacities and can be active,with batteries, totally passive, no batteries, memory retention with power for communication provided by inductive transfer or passive read/write with battery-backed memory. Memory module circuitry is comprised of an antenna system, decoder, memory and voltage regulator.

Figure 3-1

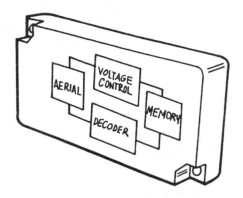

ORIENTATION & RANGE

Several considerations need to be addressed in regards to orientation and range. First, range, the distance between the memory module and the reader, will be dependent on signal strength from the memory module. High frequency, active modules will have a greater range than low frequency, passive modules. A critical question to ask is if the read and write distances are the same. In several cases, memory modules can be read at long distances but require closer memory module-transponder proximity for writing to the memory module. For best results, the memory module should be installed to pass through the strongest —largest—part of the read/write field rather than the furthest portion of the field.

For proper information exchange, a nominal distance for transmission is recommended by the manufacturer. This distance permits the system to offer maximum mechanical latitude—misalignment—and reliability. Manufacturers will also provide recommended lateral and longitudinal alignment tolerances.

Figure 3-2

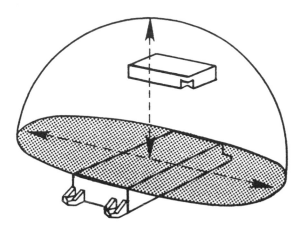

Direction of travel can also affect data transfer speed. Data transfer in direction 2 to 1 is slower than in direction 3 to 4 due to antenna placement in the memory module and their direction of travel when entering the capture field.

As the memory module enters the interrogation field, the host computer or programmable controller addresses the reader/writer via a simple protocol to either read information or write new data at a specific location in the memory module. The data transfer takes place via radio frequency technology.

Figure 3-3

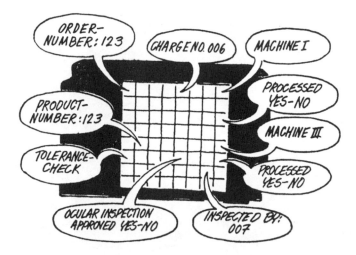

Read/write units transmit to and collect information from the memory modules. This communication can take place while the memory module is stationary or moving. Commercially available reader/writers communicate with the controlling computer system via parallel or serial interface.

Read/write units are also available which incorporates a mini-processor and can function as a limited local system controller. This type of unit makes it possible to process data, make and implement decisions based on information carried in the memory module. Advantage: fast, simple decisions made without burdening the controlling computer.

Figure 3-4

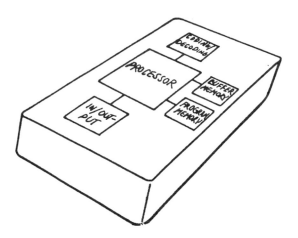

A summary of manufacturers of transaction and portable data-base equipment is shown in Table 3-1.

Table 3-1

TRANSACTION and
PORTABLE DATA-BASE
RADIO FREQUENCY EQUIPMENT
MANUFACTURERS

Company Name	Tag Name	Operating Frequency(s)	Memory Capacities (Bytes)	Read Distance (Maximum)	Read Speed	Write Distance (Maximum)	Write Speed	Active/ Passive	Read/Writes and Data Retention	Comments
Allen- Bradley	2750- TFAW2K	915 & 1830 MHz	2k 8k	60 inches	20ms/ Byte	11 inches		Active	10,000 cycles	
Balogh	OMA,OMB OMC	1.5 MHz	64 2k	3 inches	*650ms/ 64 Bytes		650ms/ 64 Bytes	Passive R/W Active Memory		*Read speed - stationary tag
Escort Memory Systems	2000	170 & 180 kHz	2k 8k	24 inches	100ms/ 16 Bytes	6 inches		Active	2,000,000 cycles	Technology Licensee - Namco
NDC Automation **Statec	MROS	30, 100 & 720 kHz	64 2k	15 mm	10ms/ Byte	15 mm	*20ms/ Byte	Passive (Hi Temperature Available)	100,000,000,000 cycles 10 yr retention	*Write includes a read ** Patented
Proximity Systems Inc	N/A	170 MHz	28k 124k	100 feet	N/A	100 feet	N/A	Active	N/A	
Redar	ASDIC-H	434 MHz & 455 kHz*	2k	16 feet	280ms/ 64 Bytes	16 feet	280ms/ 64 Bytes	Active (Hi Temperature Available)	N/A	*European Frequencies

Technology

FACTORS TO CONSIDER

Integration of RF automatic identification and portable data-base systems have several unique factors to consider:

- Frequency
- Orientation & Range
- Capacity
- Durability & Life Cycle
- Reliability
- Programmability

Frequency _____

As you can see from Table 3-1, operation of portable data-base equipment is divided between high and low frequencies.

High frequency equipment operates above 500 MHz, with some microwave RF systems emitting signals as high as 2.5 GHz.

Frequencies above 500 MHz have characteristics similar to visible light and propagate wave patterns in a narrow beam. Because of this type pattern, memory module orientation while in the capture window is critical. The plus side is long read/write distances. High frequency signal energy is also susceptible to absorption by liquids and grease thus introducing signal attenuation and distortion in an industrial environment.

Multi-pathing (" cross-talk") due to reflection of the emitted signal off surrounding metal is another point to evaluate and is usually a limitation found in very high frequency transmissions.

On the other hand, high frequency equipment has the ability transfer data fast.

Low frequency equipment, below 500 MHz, operates similar to your AM radio. This type of equipment is less sensitive to memory module-reader orientation. It, in essence, becomes omnidirectional.

Low frequency equipment provides better penetration of objects between the reader/writer and the memory module.

Figure3-5

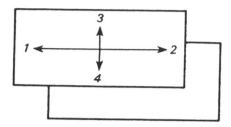

Plane and spatial angles—pitch, yaw, and roll)—will also affect module performance. Maximum values are given by the manufacturer and one must be cautious in adding these values together. Problems will occur if memory modules are used at maximum values.

Figure 3-6

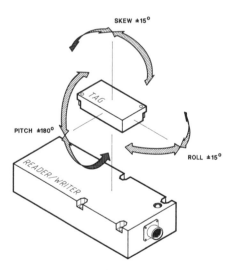

Finally, unit spacing to prevent RF interference. Based on the signal pattern emitted by the read/write unit, one must determine how close together memory modules can be counted.

Figure 3-7

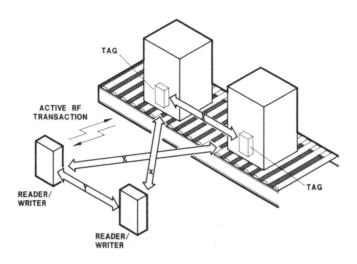

CAPACITY

Generally, one of the first questions asked is "how much data will it hold?" Transaction and portable database modules cover a span from 64 bytes to over 100K bytes of information. In what format this information is stored and how much memory you will need varies from manufacturer to manufacturer. Data can be stored in binary, hexadecimal, binary-coded decimal (BCD), and ASCII formats.

A key feature to look for is random access of bits and bytes within the memory structure. Less actual memory will be required if you can directly access a bit or byte and work off of or onto this information.

Memory space can also have restricted areas used by the manufacturer for configuration and maintenance settings. Configuration information is stored within a separate non-volatile memory area with bit-settings representing function of the memory module and/or reader/writer.

DURABILITY & LIFE CYCLE

How long is the equipment expected to function in your environment? Do you have a harsh factory or outdoor environment or a forgiving indoor or office-type environment? Check for temperature ranges, shock/vibration, fluids, and electrical noise.

BATTERY vs. NO BATTERY

Life durations of batteries in CMOS memory structures depends on the battery electrical capacity (Qc) calculation divided by the consumption (Ic).

$$t = \frac{Qe}{Ic}$$

The difficulty in estimating battery life resides in the fact of knowing how the system degrades itself.

The CMOS-RAM is subject to two types of operation—work and rest—with the current draw of each being substantially different. Life duration characteristics are very often based on the current draw while the CMOS-RAM is at rest. Dispersion of this characteristic is large and will vary with temperature.

Figure 3-8

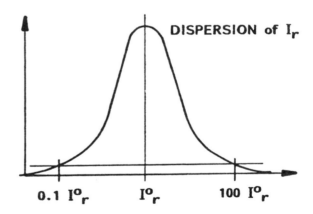

Immunity from parasites and mean-time-between-failures is an additional point to consider. The CMOS-RAM is constantly supplied with power, therefore it is polarized. This polarization leaves the memory more vulnerable to stray electronic noise than a structure without battery back-up.

Figure 3-9

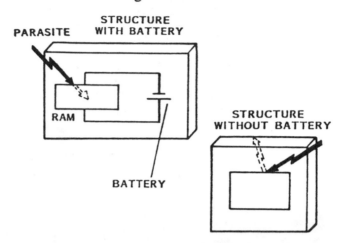

The stability of lithium batteries is obtained by their low internal power consumption. Positive ions are heavier than negative electric charges. Under a centrifugal effect—loads caused by shock and vibration—this equilibrium is reduced causing internal power leakage.

This equilibrium is also reduced by thermal effects.

Figure 3-10

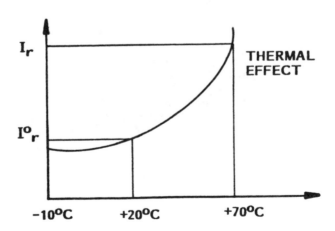

There is difficulty in defining the life duration characteristic from the calculation of components Qe and Ic whose behavior over time is unknown. Because of this, battery operated memories in industrial environments should be questioned.

Figure 3-11

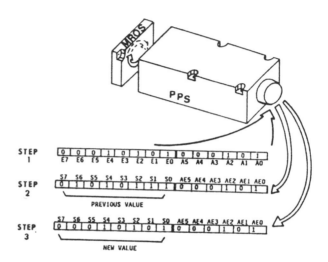

RELIABILITY & PROGRAMMABILITY

How certain are we that each memory module has been written to or read from correctly? Data integrity is affected by the previously mentioned factors of orientation, range, separation of modules, and environment plus error management.

Error checking is accomplished different ways by the various manufacturers. First, check sums may be used. Check sums encode each byte into two Manchester-type bytes. If check sums equal, data is correct. This can eliminate the need for multiple reads.

Another method incorporates a cyclical redundancy check (CRC) of several words. CRC is obtained by adding the contents of fields N to N3, and subtracting the lower 8 bits of the sum from 256.

Finally, data echo is used for error management. This is accomplished, for example, in a write cycle as follows:

*Write the value 15H to the addressed word #5.

Step 1.

Place the values 15H on the data inputs (E0 to E7) and)5H on the address inputs (A0 to A5) at the same time. The PPS transfers these values into the MROS. (Note that the contents of word 5 are NOT modified at this time).

Step 2.

The MROS returns an echo of the address (05H) and the current value of the word at that address. The PPS displays this word value on the outputs S0 to S7, and the address echo on AE0 to AE5.

Step 3.

The address echo is compared with the desired address to confirm that the MROS is accessing the correct word, in this case, word 5. At this moment, writing can be ordered—with the write signal. The MROS then returns the echo of the specified address along with the new value of the addressed word. This address echo and word value are displayed on the PPS outputs AE0 to AE5, and S0 to S7, respectively.

A common question arises ... "How fast can the tag go?" The amount of time (in seconds) that a tag is present in the RF field is called the Capture Time. This is directly related to the size—in mm, inches, feet—of the RF field, called the Capture Window, and the rate—in feet/minutes—that the tag passes through this field, called the Tag Speed. Any one of these parameters can be calculated as a function of the other two, using the equation:

$$\text{Capture Time} = \frac{(60 * \text{Capture Window})}{\text{Tag Speed}}$$

There is a minimum period of time required for each type of tag data transfer. This is the amount of time required to perform the given operation under ideal conditions, with no retries.

One manufacturer calculates this speed as follows:

Minimum transfer times for the various operation types are:

Operation Type	Min. Transfer Time
Decode R/O tag	200 msec.
1-32 byte R/W transfer	20 msec.
Program EPROM tag	840 msec.

To provide for the less-than-ideal conditions which usually exist, a Capture Time 5 to 10 times greater than the minimum transfer time is typically used when these calculations are performed. For large R/W transfers (1K bytes or more), a factor of 2 to 3 is adequate.

Another offers the following: data is transferred between the reader/writer and tag in 16-byte blocks. The amount of time typically needed to complete a transaction between the reader/writer and a tag depends on the number of blocks to bet transferred. Each 16-byte block transferred takes approximately 100 milliseconds, with an additional 100 milliseconds at the beginning of the transaction due to tag start-up time. Read operations and write operations take the same amount of time. Read operations and write operations take the same amount of time. A transaction could take longer if the tag is in the marginal region of the read/write field. The equation for total time per operation is then:

Operation time in secs. = (# bytes to transfer/16) * 0.1) + 0.1

For example, a 1 to 16 byte transaction would typically be completed in 200 milliseconds, 17 to 32 bytes in 300 milliseconds, etc.

The maximum speed the tag can travel past the reader/writer and be read from or written to is given by the formula:

Long Range Mode:
$$\text{Max.Speed(ft/min)} = 60 * \sqrt{\frac{4-(\text{Range})^2}{\text{Operation Time}}}$$

Short Range Mode:
$$\text{Max.Speed(ft/min)} = 60 * \sqrt{\frac{0.6-(\text{range})^2}{\text{Operation Time}}}$$

Range—the closest distance the tag passes to the reader/writer in feet.

Operation Time—The time in seconds as determined in the section above.

For example, if an operation calls for reading 128 bytes from a tag in long range mode set up to pass one foot from the reader/writer, the equation would yield the following

$$\text{Max.Speed(ft/min)} = 60 * \sqrt{\frac{4-(1)^2}{(128/16)*0.1+.1}}$$

APPLICATION

All this technical "stuff" is interesting, but how do I apply these transaction and portable database modules in my manufacturing facility?

Here's an example:

A completely unmanned, palletized, non-synchronous system simultaneously assembles 200 aluminum oil pumps and front cover assemblies per hour at a Canadian automobile plant, thanks in large part to radio frequency Escort memory Reader/Writers and data-carrying memory modules used in pallet and part identification and staging, and manufactured by NDC Automation, Inc., Charlotte, North Carolina.

The two produce assembly system designed by Pick-O-Matic Systems, a Sterling Heights, Michigan, Division of Crane Company, measures 12' x 80' and utilizes memory modules which are mechanically mounted on the pallets.

There are 50 pallets, each 2' x 1', which carry a combined weight of 3 lbs. for the two assemblies. A total of 17 different components are assembled into the two units during the production cycle.

Eight Pick-O-Matic parts handling robots are interspersed among the system's 23 stations for rapid parts insertion and loading/unloading. The pallets, with their database memory modules attached, allow the Escort memory to constantly track and monitor the location and status of the parts being assembled.

Importantly, Escort memory provide the assembly system with the capability of monitoring the build status of both parts on each pallet. This eliminates double cycling of pallets which are carrying one rejected and one accepted part; and ensures that each of the systems' 23 stations are running at optimum capacity.

A major hurdle facing Pick-O-Matic engineers in designing the system occurred when a pallet contained one acceptable part and one that had been rejected during the manufacturing process. Escort memory, with pallet/part identification, provided the solution. Here's how:

The initial system design incorporated mechanical flags on the pallets which differentiated the process status of each part. The flat system would only determine if a part was in "accept" or "reject" status, but could not determine at which point in the process the part was rejected.

When one part one the pallet was a reject, and the other was good, both parts would have to be stripped from the pallet to prevent further assembly of the reject part.

Further, the stage of process could not be identified when the accepted part was reintroduced into the system. Two new parts would have to be placed onto the pallet and reintroduced into the system, causing loss of production.

Escort memory modules provided the means to monitor the build status at each location. Now, when a pallet incurs a reject part, only the reject is removed at an unload station, and a corresponding new part for assembly is added.

The pallet is then reintroduced into the system. Processes are performed only on the new part until the pallet reaches a station where the memory module is read and it is determined that the current station process needs to be performed on both parts.

A second critical hurdle facing Pick-o-Matic and NDC engineers was how to eliminate a "bottleneck" that would prevent the system from achieving the required production throughput rate of 18 seconds per completed assembly.

Each pallet required a 25-second cycle time in leak test operation to process both parts of the assembly. It was thus necessary to design into the system duplicate leak test stations —numbers 5 and 6—to achieve throughput.

Initially the system was designed with a mechanical pallet flag system for differentiating pallets entering station #5—low flag—or station #6 —high flag. The "bottleneck" became apparent when it was realized that the flag system could cause a 48-second cycle to occur when a pallet entered station #5 and clamped up, while at the same time station #6 was just releasing a process-completed pallet.

Station #6 would then be empty while station #5 was performing its test functions, thus preventing required throughput.

The addition of Escort memory Modules to each pallet solved the problem by allowing each pallet its own personal identification number. Each pallet can now be read and identified by an Escort, memory Reader/Writer as it enters station #5. The controls logic of the system is then searched to determine a status: either station #6 is at empty state or it is within an established critical set-point state of its process cycle completion.

If one of these two status situations is present, then the pallet being considered at station #5 is either released or held momentarily and then released into station #6.

The result with Escort memory allows the Pick-o-Matic system to keep both stations at optimal up-time while also achieving the necessary 18 second cycle production throughput of assemblies at all 23 stations..

Several other advantages to the system were discovered and cited by Pick-O-Matic engineers as they began to realize and utilize the full capabilities of NDC Automation's Escort memory.

For example, the ability to multiplex or combine reader/writers onto common I/O cards saved considerable dollars, conserved critical remaining I/O points, and allowed easier controls programming with minimal fault points.

System debug was also made easier by being able to identify different stages of the process at each station.

Pick-O-Matic engineers estimate the new system has increased production of oil pumps and front cover assemblies (which are mated at the end of the manufacturing operation) by 20 percent over previous methods which used manpower intermingled with automatic machine and transfer lines.

The new system is also furnishing a superior product, assuring almost flawlessly assembled units coming off the line.

The Automatic Identification Manufacturers (AIM) Association's brochure on RFID summarizes this technology well:

"RF identification systems play an increasingly important role in production, inventory, and asset and resource management. Their ability reliably capture accurate information on goods-in-process has made a significant contribution to the performance of manufacturing systems. Further, their ability to acquire and transfer data in real time enhances the performance of contemporary machine-control and material-handling systems. RF systems add a new dimension to automatic identification by providing solid solutions to the environmental challenges that alternative systems have been unable to meet. Tag costs, although somewhat higher than those for bar codes or OCR labels, may be easily justifiable for more challenging applications. RF reader costs are comparable to those for other automatic readers."

"RF systems are an important new tool for the system designer. Like any tool, they must be evaluated on the basis of cost and the projected benefits. Like some tools, they may well be the only solution for a given application. We expect to see them in an increasing number and variety of systems as designers become familiar with the capabilities and potential for RF identification technology."

REFERENCES

1. NDC-Statec "Right Time" Brochure
2. NDC Introduction Course class notes,
    page 7-10 Appendix A; page 5, General Discussion
3. Namco Publication IM-SNII-01, Page 7, 21-23.
4. Allen-Bradley IA Specification, Page 12
5. Automatic Identification Manufacturers Association
    "Radio Frequency Identification" brochure page 3.

Comment: 3.4

The following article was intended, I believe, to attract attention through the development of controversy. It succeeded in attracting some attention, via selected quotes which were sometimes a little out of context, and a little embarrassing to a couple of the interviewees. The industry , and those interviewed, survived the tempest in a teapot. The author is not from the industry nor is he a user of the technology, so his viewpoint was not parochial nor discrimiantory and it is included here because I believe it makes a contribution to the readers' understanding and perspective.

3.4 RF TAG

Promises and Problems of RF Identification

Dennis Bathory Kitsz—ID Systems—© December, 1987

As with a single mind, 50,000 salmon make their way upstream, slipping past a huge dam. As each one sidles through a narrow plastic pipe, a U.S. Fish and Wildlife Service computer records its passage. I almost imagine their names—Arty Salmon, Josie Salmon, Louis Salmon, Nancy Salmon, Norm Salmon, Sam Salmon—as they swim, no longer anonymously, up their upstream destination.

The technology that removes a salmon's anonymity is radio frequency identification (RF/ID). This same technology can guide a robot through a factory maze to its proper tools; it can verify personnel security clearances—even changing the information as they pass through a door; it can feed a dairy cow the proper amount and update its medical records and milking history; or it can track containers in an environment filled with harsh chemicals, sprays, grease, and shifting temperatures.

At the same time, RF/ID is a soap opera of conflicting claims, incompatibilities, and "almosts."

What is RF Identification?

Radio frequency identification combines new and old techniques to offer miniature, hands-free tracking of products, animals, equipment, and personnel. The old technique involves several varieties of radio transmission, and the new one is computer memory and digital processor control. A miniature receiver/transmitter and accompanying computer memory—a combination called a transponder—can be encapsulated in a card or tag sometimes as small as a grain of rice.

When a transponder comes within the range of a reading station, it is activated. The characters stored in memory are transmitted to the reading station, which then decides how to use the data; this is "reading" the data. In the simplest situation—the spawning salmon, for example—an identification number is stored in a computer file. More complex uses involve changing and updating thousands of bits of information in the transponder's memory—"writing" data. The entire process takes place in a fleeting few seconds as the transponder passes within reader's range. Both range and reliability of data are key problems in RF/ID; additional questions include battery reliability—an issue in serious dispute—and standardization.

Why Choose RF/ID?

RF identification has its esoteric applications such as salmon counting, but it succeeds best as an alternative to other, more familiar technologies such as magnetic cards or bar codes. Dan Gallivan of Cotag International believes potential uses of RF systems should consider environment. He advises, "If they're looking at a bar code system, and they have environmental problems—dirt, grease, heat, that type of thing—that degrade the reading to, say 95 percent, then they should go to RF/ID."

RF/ID is also being increasingly used in areas where robotics are important—tool identification, for example. In messy assembly plants, optical methods such as bar codes are unusable. Hardware solutions, such as specially keyed equipment, require retooling for every change. RF identification tags, however, are fully encapsulated and protected from the environmental rigors of grease, chemicals, paints, sprays, and scratches. The robot broadcasts the computer equivalent of "which tool are you?" The tool responds, "I am this tool."

Tom Payne, Vice President of Sales at Identification devices, Inc., describes his system: "There's a small transponder in the tool holder so the machine can verify the tool when they set it. They can register which cutting apparatus is in the tool holder and can maintain central control, delivering those tools to a work cell, to automated machines. The machine, as it reaches out into a magazine and gets the tool, will verify that the tool it got is the tool it thinks it got before it starts doing cutting work. These machines are making parts for jet engines—the application has to do with control and short setup times. These are relatively low runs of high-value items; the set-up time becomes very critical."

Another industrial use is robot guidance. By installing a complete antenna grid in the floor of a plant, industrial robots of all sorts can be tracked and guided to their tasks. Pallets of raw materials can also be identified on their way to the factory floor through what Payne calls an RF "license plate." Again, contaminants such as grease and cutting fluids do not affect the system.

The growth of agribusiness has made efficient farming essential, and the dairy industry is no exception. Eureka Systems manufacturers a tag the size of a domino, complete with a microprocessor and 115 characters of memory, powered by a lithium battery. The unit weighs about an ounce. Don Elliott, Engineering Manager at Eureka, outlines its use in "animal ID, particularly cows and pigs. Our tag would either be an ear tag in the case of a pig—that's the best place to put in on a pig—or a collar for cows. Animal identification is for automatic feeding, weighing, and milking. They're also used in animal research that involves cats, chickens, etc." Elliott noted that the University of Illinois uses Eureka's tag; he pointed out that the tags' battery-powered design gives them over a 3-ft range, a respectable distance for current RF/ID systems.

Like Eureka, IDI offers an animal tag. Tom Payne pointed out that at feeding time the animals "present themselves to the feeding trough, and the readers said, 'Oh yes, this is such-and-such an animal, and it's entitled to a certain ration.' They keep a measurement and feed an animal scientifically and automatically with controlled portions."

IDI's salmon scheme, which can track a specific fish in migration studies, uses an implantable transponder about the size of a grain of rice. Tom Payne describes the system: "In the last two years the U.S. Fish and Wildlife Service has released about 50,000 fingerling salmon into the Columbia River for migration studies. They have a holding tank at the dams, where they scoop off the fish so they don't go through the generators. They force the fish to go through the tanks, flushing the water out of the flues at a fixed rate of 14 feet.. per second. We have a plastic PVC pipe—a sewer-type pipe—with antennas wrapped around it, and the fish in the water are flushed through the pipes. We build readers at the dams on those rivers that read the fish and clock them as they go through. The range is about 14 inches; it's really 7 inches, but it's 14 because the antenna is wrapped around the circumference of the pipe."

Security access and control is another hands-free application. Cotag's Dan Gallivan describes it. "In personnel tagging, you have tags on individuals. They walk up to a door or portal where they are automatically identified. Either they get access or they don't." The credit-card-size tags can be printed or have photographic identification as well.

Among other uses are automatic vehicle identification systems offered by Eureka, NDC's communication with monorail trains "on the fly," automated storage-and-retrieval systems, and electronic signposts of all kinds.

Types of RF/ID Systems

RF identification systems are manufactured in several combinations of configurations, including battery-powered and non-battery devices; low-frequency and microwave systems; read-only and read-write memories; intelligent and simple tags; AM, FM, and non-carrier transmission methods; and in many varieties of sizes, shapes, weights, and packages.

The choices available and compromises needed for each choice can be daunting. Among the major decisions:

- Low-frequency vs. microwave: low-frequency systems—under 1 MHz—are inexpensive, safe, and license-free. They also suffer from short ranges—about 5 feet maximum—and slower data transmission. Microwave systems—over 900 MHz—increases the distance to 20 feet or more and offer fast data transfer, but may require FCC site licensing and, when incorrectly used, can be dangerous for nearby personnel.

- Battery-powered vs. non-battery: Battery-powered systems make possible greater transmission range, higher memory capabilities, and microprocessor intelligence. They are also more expensive and may suffer from battery failure and lost data. Non-battery systems are powered—"excited"—by the radio frequency energy of the read-station itself, so they are small, simple, and failure-free. On the other hand, the range is short—some as little as 2mm, memory capability is usually small, and they are rendered instantly inactive when they leave the energizing field.

- Read-only vs. read-write: Read-only tags are basically electronic labels with permanent information and are comparatively inexpensive. Read-write tags can have their information changed either by the incoming tansmission or by the connection to a special tag writer; some of these tags cost nearly $200 each.

- Simple vs. intelligent: Simple tags contain information only, whereas intelligent tags not only update information, but also check battery life and data validity, and can be programmed for a variety of computational tasks.

- Transmission methods: Depending on the electrical noise in the local environment, more expensive noise-resistant transmission techniques must be used. AM is least expensive and most susceptible to noise. Some non-carrier methods are successful in rejecting noise. Expensive FM and microwave methods result in clean data. Systems also include error-checking and correcting and multiple transmission methods to improve the accuracy of transmitted information.

Most of these choices are decided by the application, but unfortunately, it is not a true mix-and-match situation. For example, the reliability of the non-battery system is hard to deny. According to Peter Paijens, U.S. marketing and sales support manager at NEDAP, "With a non-battery tag, you are absolutely sure that the thing will work almost forever. Tags found on garbage fills were still working. Farmers throw away animal ID tags and forget about them. Five or six years later they find a tag, and it's still working. I wouldn't say that about a battery tag."

Paijens expressed serious concern about damage to the environment from the highly toxic lithium cells used in battery-powered tags. "The environmental issue is becoming more and more a problem, especially if you are selling large numbers—millions, which we expect in the future—or if you are looking at implantable devices. They're getting into the environment. We've made calculations for that, and it's considerable."

On the other hand, John Chilton of Escort Memory Systems, Inc., is firmly behind the battery-powered tag. He dismissed the complaints, saying, "The best batteries are polycarbon monofluoride, which are used in consumer applications now. They're just as reliable as you can get. The problems were solved because Kodak spent millions of dollars to get them reliable enough and to get permission to sell them over-the-counter in their disc cameras. Now they can put them in bubble packs and sell them in the drugstore. They're just not a big deal anymore." He continued, "How many times have we put a battery in a calculator or a radio and have it not work? It's just a reliable thing." Will they die unexpectedly in the tag? "That's just ridiculous."

Mark Sekerak, RF/ID product manager for Namco Controls, agrees. "We have actually done better than we predicted. We predicted that we would get a life of ten years on lithium batteries, or three to four million read/write operations. We have actually been seeing four to five million operations." How does Namco prevent problems from dying batteries? "We have a low-battery indicator, and we interrogate the tag for battery status. We can keep an ongoing check on the number of cycles as well." The Namco tag transmits its serial number, its date of manufacture, its size, the cycles completed, the number of reads and writes, and the battery indicator.

Escort's Chilton challenges the reliability of the RF-powered—non-battery—tags. "The real problem is that when you take the tag out of the field, it turns off right then. It doesn't do its housekeeping first, finish its job, and then put itself on standby. It's like pulling the plug on the computer: whatever is happening just stops. So if the tags pass out of the field, it's write, write, write, write, write—die! And it's dead."

Ron Ames, RF committee chairman for AIM, explained that range is the major reason for adopting high-frequency—microwave—RF/ID systems. In one application, "They were doing it with trucks; they were getting 30 or 40 feet. It is a question of how much power you put out in an industrial environment. The readers at higher power levels have to have size licenses." But the units can be optimally reliable. "I have had two tags in my hand and have passed them in front of a single reader. One of them was battery powered and one wasn't, and at about 10 or 15 feet, both tags were read, even in a lab where there were several other readers operating—a pretty cruddy environment—and still they seemed to read reliably."

Gallivan and Chilton both expressed concern about microwave RF/ID. "It's more directionalized, and you're going to run into problems with shielding, incomplete codes, or reading more than one tag in the zone at the same time, said Gallivan. "With low frequency, you can actually read around metal." Chilton described microwave systems as having cold spots: "The reader/writer's RF field contains nulls, where the tag is completely unreadable."

"They'll Look at You and Laugh"

How strenuous these disagreements can get is reflected in my interview with David Wm. Draxler, marketing manager at NDC Systems, Inc. Draxler claims that NDC and its parent company, Statec of Paris, originated RF/ID technology. Their non-battery tags contain considerable data—up to 2000 characters— operate on three separate frequencies—one to transmit, one to receive, and one to excite the tag—and use patented memory technology. This technology, the heart of NDC's tag with its claimed 17-year memory retention life, Draxler firmly declined to discuss, saying only, "It's potential wells, which doesn't mean much to anybody. Call it blue smoke and mirrors."

Curious about a technology that claimed an astounding 1 in 10^{33} error rate—one error in a billion trillion trillion trillion cells—I turned first to an NDC competitor. Though declining to be identified, he dismissed NDC's claims of unlimited write cycles and high reliability. "My reaction is," he said, "if someone's got a part like that and there's a

Applications

Although all manufacturers attempt to describe the sorts of things that their products would be useful for, there is no consistency in the methodology of these descriptions.

Analysis of these descriptions show that some of them categorize the applications, at least part of the time, by the objects being identified. In other instances, categorization is on the basis of abstractions about the objects. Still others are categorized by a more general function or process that the components were part of.

It is useful in trying to communicate about these applications to identify related applications within the same category of use distinction. In other words, to make comparisons between the objects that are being identified from system to system, application to application, and even industry to industry.

It is also obvious that in some applications, although the objects are different, the abstractions are the same. An example of abstractions would be STATUS, such as empty or full. This attribute can be applied to a broad range of objects in a subcategory which might be called, CONTAINERS. Another such abstraction would be LOCATION. This could be applied to virtually all objects.

The third category, which is FUNCTION, can encompass the previous two. For example, the function INFORMATION CAPTURE can relate to the object and abstractions about the object. For example, all vehicles weighing more than 10,000 lbs. Another example of function is TRANSACTIONS. It subsumes an object which is a person and an abstraction which is value and could be indicating the function of making a debit or a credit to an account.

In the following table, I have listed categories of objects which appear to be gaining acceptance as the major set of categories which will be used for communicating within the industry. There is less awareness and acceptance of the categories listed within ABSTRACTIONS and I am certain that the list is not complete or refined. This is also the case under the category, FUNCTION.

I hope by making the you aware of these categories, it will contribute to the development of some standard terminology which will make communication easier between manufacturers and their customers, their investors, and related industries.

Table 4-1

OBJECTS	ABSTRACTIONS	FUNCTIONS
People	Status (1/0)	Entitlement (Access/Use)
Things	Time	Information Capture
Places	Rank	Routing/Tracking
Vehicles	Quantity	Storage/Retrieval
Animals	Measure/Metric	Process Control
Money/Value	Location/Route	Transactions

Having addressed the subject of categorization let's look at some real examples of ways this technology is currently being used. In this chapter there are examples of some of the pioneer applications as well as some of the newest.

Comment: 4.1

One of the largest applications of RF/ID since 1973 has been Security or Access Control. The following article describes one of the many thousand world-wide installations of RF/ID Access Control Systems. RF/ID offers substantial benefits in both security and in convenience.

A Sample of Low-Frequency RF/ID Systems

Table 3-2

Company	Tag Type	Power Cell	User Memory	Usable Range	Frequency (KHz)
Cotag	R/W	Yes	64 bits	1.5-6 ft.	66, 132
Eureka	R/W	Yes	115 chars	39 in.	66, 132
NDC	R/W	No	64-2K chars	5/8-2 in.	30, 100, 720
Namco	R/W	Yes	8-32K chars	6 or 24 in.	178
Namco	Rd	No	16 chars	12 in.	37, 148
Escort	R/W	Yes	8-32K chars	6 or 24 in.	178
IDI	Rd	No	35 bits	7 in.	40/50, 400
NEDAP	Rd	No	32 bits	28 in.	100-200
NEDAP	R/W	No	128 chars	2 mm	N/A

Table 3-3

Company	Error Rate	Reader Cost	Tag Cost
Cotag	10-28	$200	$15
Eureka	99%+*	$1100	$20-35
NDC	10-33	$1500	$100
Namco	10-14	$1800-2400	$150-175
Namco	10-14	N/A	$25-30
Escort	99%+*	N/A	$150-175
IDI	10-6	$500-1000	$5-20
Nedap	N/A	$1500-2500	N/A

*Accuracy rate rather than error rate. N/A = Not Available.

Dennis Bathory Kitsz is a musician, writer, and publisher. He is currently exploring the creative possibilities of keyless data entry.
ID Systems—December, 1987

billion dollar market out there for EEPROM—electrically erasable read-only memory—why aren't they marketing it? You're not going to put that in RF tags."

I spoke next with NDC's Larry Freeman, who admitted that it was an older technology reapplied to a new situation. "The world's been excited for the last ten years about megabits on a chip, more and more memory on a chip, lower and lower cost, faster and faster speed. And here's a guy in France who says, 'Wait a minute! I don't need all this. I need the exact opposite.' Use large memory cells, realize that a couple of words is really all you need, and access time in microseconds is not important—this was the method that he came up with to design our memory chip." As to why the chip is not commercially available, Freeman said, "Try to sell a memory chip for $1 a word; they'll look at you and laugh." Present-day memory chips cost less than 1/10 of a cent per word.

Ron Ames, RF/ID committee chairman for AIM, refereed these conflicting comments with his speculations. "He may be using some more primitive version of EPROM technology. No one should worry about it. If it works, fine; if it doesn't fine." But Ames was concerned in another way: "If in fact, it is a strange process, they're not likely to get much volume—and therefore cost reduction—out of it." Ames worried that the proprietary technology would leave users tied to NDC as a single source. He continued, "I would be much more concerned about error management. Since it isn't a technology that's well known, it's very difficult to predict what the error rate would be."

Solving the Problems—Now? Or When?

It becomes obvious that range is a serious issue in the application of RF/ID systems, as is the dispute between advocates of battery-powered and non-battery transponders. The question of standardization—the ability to select form among several vendors for price and value with the assurance that all the pieces will work together—has hardly been addressed.

Ron Ames states, "RF/ID technology has almost arrived. To some extent, the technology is there now. There's only one thing that keeps them back from the real plum applications, and that is standardization. Nobody's system is ever compatible with anybody else's. A lot of these are open systems, and open systems need multiple vendors with compatible stuff. As long as it's closed and you're just doing the dairy cattle on your farm coming up to get fed, it's easy. And to some extent, it's fine in the security and access control business because it gives you that much more security. But that's about where it ends."

4.1 OPEN SESAME IN THE 1980'S

Radio frequency tags and readers bolster security within The Bank of Boston's London office.

Tony Whiting—ID Systems—© March, 1989

One of America's oldest banks is using the latest British radio frequency access control technology to bolster existing internal security at its United Kingdom headquarters in London. The First National Bank of Boston, which specializes in providing financial services for corporate and high net worth clients, has installed a hands-free system from Cambridge-based Cotag International.

"It's very important for us to be able to control access to the bank's sensitive areas, such as the dealing and computer sections, and equally crucial that the system we use offers effective, reliable, hands-free operation," said Bob Barber, senior manager of operations and administration.

"The Cotag system supplements our conventional security arrangements perfectly. We control who comes into the building, and Cotag says where they can go."

Search for Security

The First National Bank of Boston was prompted to reconsider security upon relocation to a building that was the subject of a major refurbishment. The Bank had previously used both digital and card access systems and had experienced shortcomings with both access control methods. The problems were further compounded by the fact that the Bank was only planning to occupy six of the eight floors.

The Bank, after looking at sites using the latest generation of systems available from former suppliers, was not convinced that they were being offered the most suitable system. The Bank decided to investigate

Cotag's system when Nico Construction, the main contractors for the building refurbishment, recommended the system and offered their existing clients as referrals. Representatives of Bank of Boston visited several sites, including the London headquarters of a major European Bank, where a Cotag system controlled access to two buildings.

Backed by the British government as well as key U.K. investors, Cotag's RF access control offers advantages over conventional keypad and card equipment and short-range proximity systems—all of which require the user to find the card and show it to the reader.

With the RF system, tags can be carried in a user's pocket, handbag, or briefcase; clipped onto a keyring or neck chain; or incorporated in a photo ID tag. The system does not rely on contact between the tag and reader, and tags are read quickly at the door.

Because the system offers user convenience, the staff needed only a brief introduction to its use. Personnel simply wear the tag and approach the door as normal—no action is required. When an unauthorized person approaches a door, access is denied, and if required, the event can be recorded or an alarm sounded.

All employees have tags, and each tag is individually coded—9.2 million, million, million different codes are available. Each tag has its own access level, i.e., it can go through certain doors at certain times. Furthermore, each tag is security coded for Bank of Boston only and only authorizes access at this particular site.

Clients and visitors to the bank must report to the reception area. From there they are accompanied to the appropriate area or staff member.

Although restricting the access of certain personnel and visitors to some areas was the basis of the installation brief, the Cotag system, when connected to a PC, will also enable bank security officers to monitor staff movement.

"By linking the Cotag reading heads to a PC, we will be able to monitor exactly when people come and go as well as to simply identify them. If an incident occurs within any part of the office, we shall be able to narrow down considerably the possibilities of who could be involved."

An Eye on Aesthetics

A feature of the prestigious Bank of Boston building on Victoria Street is its modern interior design. Aesthetics demanded that the access control system blend carefully with the decor.

"Our design team was understandably a little concerned as to how we could install a complete access control system without disturbing the aesthetics of the building too much," Barber added.

"Cotag avoided too much wringing of hands by supplying reading heads which were able to blend in unobtrusively with the office decor."

Tony Whiting is a journalist specializing in access control technology, with Paul Sowerby Associates of Harrogate, Yorkshire, England.

Comment; 4.2

The next article is about identification of objects called vehicles. The program discussed has been slow developing, however, it has recently been gathering some momentum. The company mentioned in the article, Identronix, has been purchased and integrated into the Allen-Bradley Company.

4.2 The Legals Keep on Truckin'

Scan Journal, Volume I, Number 9—© Third Quarter, 1985

Overweight trucks tear the tar out of roads and scare the tar out of automobile owners. At best they sap highway repair funds and place legal carriers at a competitive disadvantage. At worst they are a menace to everything in their path.

Catching them has traditionally been a hit-or-miss operation in many states, depending on weight station operating hours and the ability of state police to spot low-rider trailers. But something is happening in an area called the Crescent to change all that . . . HELP is on the way as part of the Crescent Project.

Department of Transportation in the Crescent, a region stretching from British Columbia in Canada south through Washington, Oregon, California, Arizona, New Mexico, Texas and then veering north to include Arkansas, have instituted a pilot program that identifies, weighs, classifies and catches illegal carriers—at 55 mph.

The Crescent Project marries automated vehicle identification (AVI), weight-in-motion (WIM) and automated vehicle classification (AVC) through the use of Heavy Vehicle Electronic License Plates (HELP), placed on the tractors of selected carriers.

HELP's are passive radio frequency (RF) transponders about the size of blackboard erasers which receive their power from the radio signal of an interrogator—transmitter/receiver—both manufactured by Identronix.

The Crescent Project system, developed by General Railway Signal (GRS) ties the RF ends to a computer and a load cell scale—no swinging weights or balance arms—from CMI Dearborn Scales.

As vehicles speed down the highway, the interrogator sniffs about at low power until it detects a transponder. Then, when the vehicle is about to pass over the scale, the interrogator boosts its output at 915 MHz to trigger the transponder. The transponder responds with the identification data at 1830 MHz. All this takes only 30 to 100 milliseconds, depending on the amount of information involved. The identification information is fed directly to a computer which registers the time of weighing, weight compliance and vehicle speed, then checks the tractor or trailer for previous violations, registration and safety inspection records. Other information categories can be added to the data base as needed.

Quite a difference from the 1 to 2 hour wait at some weigh stations.

Considering the time savings for legal operators and the reduction of road damage alone, the system should receive widespread support. However, while access to automatic vehicle identification and location data have far ranging national implications for DOTs, what about the independent and company carriers? The Crescent Project will help them improve operations by providing for:

1. Better monitoring of vehicle movement, weight, and speed;

2. Improved data on which to base maintenance schedules;

3. More timely data and enhanced control on leased vehicles;

4. Automated billing and accounting;

5. Fair competition by excluding illegals from the highways;

6. A substantial reduction in the $7,000,000,000 annual losses to equipment and cargo theft and;

7. Reduction in the bane of everyone's existence, paperwork.

For carriers to reap some of these benefits, they must have access to a state's port of entry (POE) and weigh station master data base. This addition to the Crescent Project was planned from the beginning and Lloyd Henion, Director of the Oregon Department of Transportation's (ODOT) Tracking and Weighing Program, reports the system will become open to carriers on a trial basis in the near future.

For more information on the Crescent Project, contact the DOTs in individual states. For information on application of integrated RF identification systems for railroad, mass transit, box car, corporate security, access, toll booth, container; locomotive, CAD/CAM and route selection, contact GRS, Box 600, Rochester, NY 14602. Those who would like to learn more about the basic technology used in the Crescent Project can contact Allen-Bradley, 1201 S. Second Street, Milwaukee, WI 53204.

Comment: 4.3

The next article involves the use of a single (1) bit tag that saves lives.

4.3 A Coal Mine somewhere in England...

Scan Journal, Volume 1, Number 2—© First Quarter, 1985

Deep underground, an unconscious miner rides a conveyor belt toward oblivion . . . drawing ever closer to a powerful, thundering coal crusher. . .and to a sure and horrible death. . .

Sound rather like the concluding scene from a "Perils of Pauline" episode? Perhaps. But when the British National Coal Board (NCB) decided in the 1970's to install unattended lump-breakers in its collieries, it took a serious look at the safety implications. At the request of the NCB, the Mining Research and Development

Establishment (MRDE) undertook a study of safety systems to ensure immediate unattended shut-down of the conveyer in life-threatening situations.

Requirements were stringent. The system would, of course, have to be 100% reliable. In addition, it would have to be extremely selective to avoid accidental shut-down by passing personnel; insensitive to tramp iron and large lumps of coal; able to withstand the rigors of the mine; and work even if obstructed by coal or a miner's body.

After deciding that protective fencing, infra-red or thermographic sensing and magnetic field techniques were impracticable, the MRDE concentrated on tagging individual miners.

After eliminating several passive RF systems and even a radioactive tag idea, the MDRE focused its attention on a low-frequency radio-transponder system supplied by Cotag International of Cambridge, England. A battery-powered tag was developed to meet all the MDRE's operational requirements. After extensive field testing, this system was selected by MDRE.

The system has three major components; transmitter/receivers mounted on the lump-breaker enclosures, small RF tags, and a control unit.

The transmitter/receivers broadcast continually at 132 kHz, activating any tag within range. Once activated, a tag halves the frequency and responds at 66 kHz. This 66 kHz response triggers the conveyer shut-off. To prevent accidental triggering when a miner walks within the system's eight foot range, transmitter/receivers are shielded on top and sides by a steel enclosure.

To maintain system reliability, tags are checked each time the miners enter the mine. A transmitter/receiver located in a doorway reads the tag, verifies that it transmits a readable code and that the battery level is adequate. The tag transmits only when activated by the 132 kHz frequency, a feature designed to prolong battery life.

As a final precaution, tags were fitted to the miners cap lamp batteries to ensure that they would not be accidentally left behind.

The NCB has applied this same technology to vehicle location tracking by placing tags on coal cars and sensors at the turn-around points underground.

Cotag has provided hospitals with a similar system to alert personnel when patients stray off hospital grounds or into restricted areas. The programmable tags can be reused indefinitely.

For further information, contact: Cotag International (USA), 103 Springer Building, 3411 Silverside Road, Wilmington, Del. 19810, Tel. 302/651-9146, Telex: 834246; or Cotag International Ltd., Mercers Row, Cambridge CB5 BEX, England, Tel. 0223-321535, Telex: 817846.

Comment: 4.4

The following article identifies the material used to package the implantable transponder as polypropylene. This material was found to be unsuitable for the extended life needed for many animal ID applications. Glass is currently the material of choice for these tags.

4.4 Sorry, I thought you were someone else.....

Scan Journal, Volume II, Number I—© Fourth Quarter, 1985

Animals that look alike are a problem for research and other projects dealing with numbers of the same species of animals. Imagine trying to sort out a dinner party of penguins, or identify a particular salmon swimming up the Columbia River to spawn. And heaven knows how difficult it is to get well-enough acquainted with an individual starfish to be able to pick it out from an aquarium full of its brethren.

Still, positive identification of animals is important for experiments, breeding, census and farm record management. The standard methods of labeling animals are tagging or tattooing but these methods are less than perfect. First, tags can be damaged, lost or tampered with which means data integrity is limited. Second, the information from the tag must be manually entered into the information system, leaving the barn door open for errors. Tattooing horse, cow and dog lips provides positive identification but it requires manual inspection and verification.

But there is an alternative, as you knew there would be. Using a standard veterinary implanter with a 12-gauge needle, it's possible to introduce a tiny, 0.083 by 0.39 inch—2.1 by 9.9 mm—numerically programmed radio transponder under an animal's skin, into a muscle or body cavity. When a 400 kHz hand wand interrogator comes within three inches—76.2 mm—of the transponder, it answers back with its number, 35 data and 5 parity bits, at an RF basso profundo of 40 to 50 kHz.

The transponder's electronics are encapsulated in biologically inert polypropylene; extensive clinical testing earned it USP Class VI approval and there are no known side effects to the animal.

Energy to power the transponder is supplied by the interrogator signal and is a major reason implant life exceeds that of the animal being electronically tagged.

Fingerling salmon are being tagged with implants to facilitate study of migration patterns. Six-inch diameter tubes mounted with interrogators are placed in the salmon streams. Individual fish are automatically identified when they swim upstream through these tubes.

Zoological parks, breeders, research centers and locations where animals imported into the United States are quarantined have successfully used the identification system in ruminants, seals, birds, horses, swine, dogs, cats and the full range of laboratory animals.

A number of other uses have been found for the small transponders, which cost around $8.00 each in large quantities. They can be discreetly planted in such valuable items as jewelry, paintings, sculpture and books. The size also makes it compatible with small tooling that must be automatically identified by robots.

Identification Devices, Inc., the manufacturer, also makes larger transponders. One, a 2.25 inch—57.2 mm—tag has been attached like a bell to dairy cattle and is used to automate feed and milk production records. And some of the piglets that received a transponder in the ear just after birth will have their mature carcasses weighed and tracked through the slaughterhouse by these somewhat larger tags. Reading distances are from six to eight inches—152.4 to 203.2 mm.

For more information on these RF identification units, contact Identification Devices, Inc.

Comment; 4.5

The next article is about using RF/ID in order to grant vehicles the privilege of accessing a parking facility.

Richard Loomis is quoted as saying "Some of these gate openings are up to 30 inches wide and the chance of two or more vehicles going through them at the same time is very likely." This may be a typo since it is hard to imagine parking lot gates 30 inches wide allowing two motor vehicles to enter at the same time. I suspect it should have been 30 feet.

4.5 RF Tags

Solve Access Control Troubles at Airport Parking Lots

Automatic I.D. News—© October, 1988

The handwriting was on the wall, or more accurately, the driveway at Washington National Airport parking lots.

When parts were no longer available for the controlled access system at remote parking lots, International Business Services, Inc. (IBS) knew it was time for a change.

IBS, parking lot operator, had used a magnetic stripe card system for a fleet of shuttle buses, management and maintenance vehicles, airport fire department trucks and security police vehicles.

Special entrance/exit lanes were set aside in the lots for these vehicles to insure quick and easy, free access in and out of the lots. Pedestals along the lanes accepted the mag stripe cards from vehicle operators.

"Effective management and control of the use of these lanes is very important to our operation" says IBS Project Director Richard Loomis.

Unauthorized use of the cards or lanes directly affects lot operation— are the lots full or are vacant spaces available?— and financial status.

"If you can't keep an accurate lot-count total over a year's time, it can cost you a lot of lost revenue," says Charles Pillar, assistant manager, Revenue Control for IBS.

No middle of the road

Even before part replacement became impossible, IBS experienced problems with its mag stripe system. Either the pedestals were too high for some vehicles or too low for others. If the card readers were too close to the lanes, drivers would run them down. If they were moved back a little, then some drivers would have to get out of their vehicles to use them. A satisfactory solution seemed elusive. Also, constant maintenance was a problem.

To help make a decision on a new access control system for these lanes, IBS enlisted the services of Jeff Germunder of Cerand & Co.,—airport parking consultants—and Pete Martin, president of Automatic Access Controls,—automatic gate and vehicle access control specialists. Both recommended that IBS use an RF transponder type of automatic vehicle identification device (AVID) because the application was to control vehicles, not people.

A vehicle tag system eliminated the problems of pedestal-mounted card readers and, because there are no moving parts in the roadway sensing loop, maintenance is minimal. Also, since the tags are permanently mounted to the vehicles, the problem of misplaced or lost cards is automatically eliminated. This was especially true in trying to allow controlled, free access to the parking lots to a hundred or more emergency, fire and police vehicles, as well as assorted airport, FAA and Dept. of Transportation officials.

Deciding to use RF transponders mounted on the vehicles for access and inventory purposes was the easy part of the project. Choosing the right RF tag manufacturer, type of tag, method of mounting and the related access control hardware and software was not as simple.

"There are a lot of RF—proximity—tag manufacturers to choose from. However, the tag and reader/decoder you need for accurate vehicle identification must have a reliable reading range of at least 36 inches and be able to read at speeds of 30-40 MPH. The tags must be impervious to the environment, flexible in the methods of attachment to the vehicles, and have low susceptibility to potential interference caused by motor noise, gate operator detector loops and other RF devices," says Martin.

He wanted an RF tag that allowed the system to "read" two or more tags in the loop sensing area at the same time. "Most RF tag manufacturers can't do this at the present time," according to Martin, "but this feature is critical to an effective vehicle identification system."

"Identifying multiple tags in the same sensing field was not a potential problem for our parking lot application, but we needed a vehicle ID system that we could use at all our facility gates," says Loomis. "Some of these gate openings are up to 30 inches wide and the chance of two or more vehicles going through them at the same time is very likely. We hope to eventually expand this system to all of our facilities and didn't want to buy something that wouldn't work on all gates."

Security essential

Germunder had other concerns about which vehicle tag system to recommend. "As airport consultants, we are very concerned with security and a lost or stolen vehicle tag is a key to open airport gates."

An automatic vehicle ID tag must not only be reprogrammable in the access control computer if it is lost or moved from one vehicle to another, but should destroy its memory if it is stolen from the vehicle, says Germunder.

Cerand & Co. and Automatic Access Controls recommended a vehicle-mounted RF tag manufactured by Eureka Systems. Eureka also provided the loop sensor tuning units, tag reader/decoders and special "Wiegand" interface controls.

The tuning units enable RF tags to absolutely tune out stray RF interference with the system. In gate access systems, most of this interference is caused by the standard vehicle detector loops used in conjunction with electric gate operators or parking barriers providing either free exit, safety or closing functions. At the IBS sites, all loop detectors in AVID systems areas were operating on the exact same frequency that the RF tags and reader/decoders were communicating. The loop tuning units enable the RF tag system to operate within inches of these gate loops without interference.

Having the reader/decoder send out a Wiegand signal enables the system to be plug-in-compatible with most card reader—access control—manufacturers.

The Eureka tags and decoders are only half of the AVID system used by IBS at National Airport. The other half of the system is an access control package by Northern Computers. This consists of several N-1000 II logic control modules and P-3000 printer/programmers.

"Right now IBS is controlling and monitoring each site as a separate stand-alone system," says Ron Nunnally, president of Perimeter Access Systems, an OEM distributor of Eureka Systems and the installing dealer on this project.

The Northern Computer system enabled IBS to install and operate each new site as a stand-alone project since the N-1000 II is a distributed data base access control system with a capacity of up to 9000 tags. Each N-1000 II can control two lanes—entry-exit—or a combination of an Eureka vehicle tag reader and most any type of card reader or keypad at the same time."

Now and then

IBS soon plans to upgrade its VID system so all sites and activities can be controlled and monitored at its central offices. "At that time, we'll simply remove the local printer/programmers, add modems, and Northern's PC-PAC software and IBS will be able to completely control any or all facility gates from one of its existing PC/XT-compatible computers. By doing this we will increase the Northern Computer/Eureka capacity to 512 vehicle lanes or card readers monitoring a minimum of 32,000 vehicles and/or personnel."

Present operation starts with the RF tags mounted under shuttle buses, maintenance trucks, emergency and other authorized vehicles. The vehicles pass over a sensing loop areas as they approach a parking barrier arm gate.

The reader/decoder identifies the tag and reports to the controller. The controller verifies that the tag is valid, adds stored information about the tag and vehicle and reports to the printer.

The printer records the time and date; controller number and location; reader number; description of vehicle; and whether access was granted or denied.

The controller also signals the parking barrier gate to open and has the capacity to log onto the printer what time the gate opened, how long it stayed open and whether or not the vehicle actually went into or out of the facility.

The Northern PC software will add the capacity for time and attendance, group and limited use tags or cards, site plan graphics and extensive history and management reports.

"The system is now up and running fine. As with every automatic gate entry project there were some custom adjustments that had to be made; because, unlike door access, gate access systems are unique no matter how similar they appear. For this reason AVID systems should not be thought of as "off the shelf" types of products. Each system must be custom designed to the customer's requirements," says Martin.

The custom-designed RF system at Washington Airport parking lots solved IBS's controlled-access problems. Now, reading RF transponder signals is as easy for IBS as it was to read the handwriting on the wall.

Comment: 4.6

Having lived in the community of Coral Springs, Florida, myself, I can appreciate the dilemma addressed in the following article. It is the challenge of providing increased security without a corresponding increase in inconvenience. When the access control system becomes a significant obstacle to our normal activities we look for ways to defeat it.

Another community within the boundaries of Coral Springs, Florida, abandoned the use of an attendant-operated gate house a few years ago because it was a bottleneck in rush hour and the turnover of attendants required the residents to frequently stop and verify their identity to a new attendant. The following article shows how this problem can be solved.

This next article and the previous one both report a problem of local interference. Loop detectors used for sensing the presence of an automobile commonly use a frequency of 100 kHz which is near the frequency of the Eureka system (132 kHz). This may account for the first problem. A higher frequency for the RF/ID system or a lower one for the loop detector could reduce the problem in general.

4.6 PIONEERING PRIVACY

Exclusive Coral Springs Community uses transponders for gate security

Rex Valentine—Automatic I.D. News—© February, 1988

Residential privacy is a central consideration at Eagle Trace, an exclusive residential community in Coral Springs, Florida, known for hosting the Honda Classic golf tournament.

That's why Coral Ridge Properties, Inc., responsible for the development of Eagle Trace, sought an automatic, vehicle identification system two years ago that would control access through entrance gates yet be quick and automatic.

Dual-purpose entrance

The Eagle Trace East gate has two approach lanes and two exit lanes with a gatehouse in a center island. The approach lanes serve traffic from construction vehicles, contractor vehicles, community utility vehicles, and residents' vehicles. Therefore, the left-hand approach lane is designated as the guest entrance and the right-hand lane is reserved for the exclusive use of residents.

Coral Ridge Properties' goal was to have the resident lane operate automatically, requiring only the occasional attention of the gatehouse attendant whose main concern is to be sure that non-resident vehicles have proper authority to enter Eagle Trace.

Construction vehicle traffic includes some oversized vehicles that cover more than half the width of the entire access lane, making it impossible to physically separate the two lanes with a curb or island. This constraint makes many identification systems difficult to use. Any system that requires the driver to stop and present a credential to a reader device would require a pedestal to be placed between the construction and the resident lane, causing a safety hazard and preventing oversized vehicles from using the entrance. Furthermore, the driver is inconvenienced by having to come to a complete stop and roll down the window. Magnetic stripe card systems, Wiegand readers, and voice ID systems all require stopping.

From half-measure to whole system

While carefully considering various vehicle identification systems, an interim system was installed that involved garage door opener type radio frequency devices. The system catered to the convenience needs of the residents but had several drawbacks from a community management view.

First, vehicles could control the gate from a great distance, allowing them to speed through the gate. Second, the system could allow a distant car to inadvertently grant access to a vehicle closer to the gate. Finally, the devices permitted an undesirable condition whenever a control unit was reported stolen or missing. The security manager was forced to ignore the breach or to embark on a major project to reprogram each and every one of the existing units.

After considering several of the most advanced vehicle identification systems available, Coral Ridge Properties ordered an Eureka Electronic Vehicle Identification System manufactured by Eureka Systems, Inc., for the residents' entrance lane. The Eureka System has three major components that control the existing electric gate system and grant access only to vehicles equipped with valid Eureka Tags.

The central component of the system is the Eureka 411 Intelligent Tag mounted underneath the residents' vehicles. The tag is a radio frequency (RF) transponder that operates at very low frequency. It has 115 bytes of memory, an RF transmitter, an RF receiver and a microprocessor that supervises the tag's activities. The tag is epoxy encapsulated with an ultrasonically bonded outer jacket—all of which is about the size of a domino and can be attached to a vehicle using Eureka's magnetic tag holder or a rugged bolt-on device.

The tag's memory holds data that identifies the vehicle, the installation site—so that tags that work at other Eureka installations do not work at Eagle Trace—the time zones in which the tag is valid—24 hours a day, 365 days a year for residents—and an access level that can be used to limit the tag's access to particular gates. The tag's low frequency operation permits tag signals to pass through asphalt and to work well in areas near metal surfaces. It's 1000-character-per-second transmission speed identifies vehicles moving at significant speeds.

Another system component is the Road Loop Sensor that is installed in the pavement in the residents' entrance lane. Quite simple in construction, the sensor's function is to turn on the normally quiescent tag and relay the tag's transmission to the controller unit located in the gatehouse. The sensor installed at Eagle Trace is nine feet wide—across the driveway—and four feet long—in the direction of traffic—so that residents will have no problem crossing the sensor.

The Eureka 4300 Controller is the third component of the system. This unit houses a state-of-the-art microprocessor unit, a radio frequency unit, a gate control unit and a heavy-duty power supply unit. It is equipped with a serial communications port that can be connected to a monitor, a printer, a computer, a data link to a central system or to nothing at all—as Eagle Trace does—because the 4300 is a stand-alone unit.

A hand-held programmer is available that permits management to configure the controller to fit the needs of the installation.

Eagle Trace's manager of security found an immediate use for the hand-held programmer when a resident sold a car that was equipped with a Eureka tag. It was a simple matter to utilize the programmer to enter the car's vehicle ID number into the Contoller's Void Tag File, thereby forever denying access to that particular tag.

Installation considerations

Eureka's Engineering Manager, Don Elliott, outlined some of the installation considerations. "A decision must be made prior to installation as to where the road loop sensor will be located. Each installation will be different. If our customer wants approaching vehicles to come to a complete stop, then the sensor should be about 20 feet in front of the gate. If vehicles need to be identified at 30 mph, then the sensor better be at least 80 or 90 feet form the gate. At Eagle Trace, all positions were considered.

"We decided upon 35 to 40 feet to permit vehicles to gain access without stopping yet require them to reduce speed to 5 or 10 mph for safety near the busy gatehouse." Elliott also comments about newer cars having few good metal surfaces to which the magnetic holder might be attached. "Most cars are not a problem. We may have to develop another device for the difficult case."

The most difficult technical problem Eureka faced was intense electronic interference form the existing gate's control system. A vehicle detection loop used to close the exit gates generated so much electronic interference that the initial installation performed below Eureka's expectations. Elliott says "The noise was right in the middle of our operating frequency. In a way it was a blessing. My father, Dr. Daniel Elliott, a Ph.D. physicist, and I worked on a sensor design that would effectively cancel the interference.

"We made a lot of progress in a short time. The Eagle Trace installation now lives up to our expectations. We knew Eagle Trace developers were pleased when they decided to expand the system by ordering a second controller to be installed at the new North Gate."

The primary benefits of the Eureka system for the Eagle Trace community are enhanced privacy and convenience for the community's affluent residents. One additional benefit is not to be overlooked— savings in gatehouse attendant labor.

During the vast majority of hours, the Eagle Trace gatehouses only require one attendant on duty. First, by using the garage door openers and now by using automatic vehicle detection for gate control, Eagle Trace has avoided the additional labor expense normally needed with manual systems during peak traffic times.

In gatehouse situations like Eagle Trace, traffic is closely monitored in the exit lanes as well as in the approach lanes. In such locations, automatic vehicle identification is particularly effective in minimizing costs.

Eureka Vice President of Marketing Rex Valentine points out, "Even if we only save two hours of labor a day for peak traffic hours, our customers could realize a $5000 savings per year. Busier gates just make the savings that much greater. Such savings will help our customers recover their initial investment in Eureka's system."

Finally, in affluent communities such as Eagle Trace, each resident has an investment in the community as well as in his or her home. By being committed to state-of-the-art privacy and to the greatest convenience of its residents, Eagle Trace management enhances the investment of each of its homeowners.

Comment: 4.7

The next article points up the fact that multiple Auto ID technologies can, and frequently do, coexist in solving various combinations of needs in real world applications.

4.7 THEME PARKS PLAY WITH AUTO. ID

Doris Kilbane—Automatic I.D. News—© February, 1989

Grand Prix ride tracks race times with RF while cashiers scan employee badges and bar-coded ride rickets.

MotorWorld, a Virginia Beach theme park, is proof that automatic identification knows how to have fun.

The mini-race car theme park, one of several theme parks in Ocean Breeze Festival Theme Park, gives all family members a chance to test their driving skills. It also is a model for other theme parks in high-tech data gathering and managing. MotorWorld uses bar-coded tickets, bar-coded drivers' licenses and bar-coded discount coupons.

In addition, its Grand Prix track ride uses radio frequency technology to count laps and calculate elapsed time.

Another Auto. ID application uses bar-coded employee badges and a time-and-attendance program to gather staff attendance data for payroll.

If there's a way to improve management or operation procedures via automatic identification, ESG Companies, owners of Ocean breeze Festival, will find and use it.

The company's search for applicable automatic identification systems at the park began in 1987 as ESG Companies put final plans together for the multi-theme park.

"We had seen what computers, due to the quickness and volume of information obtainable, can do to assist management in making decisions," says Edward Garcia, Jr., chief executive officer of MotorWorld and vice president of ESG. "We looked at various systems and concluded the data obtained from zip codes of riders and by tracking the number of times a ride is ridden would assist us in time scheduling and time management of our staff. It also gives us the opportunity to plan the growth of the park based on what the people want."

With 60 acres of the 85-acres park left to develop, that means ESG Companies could create rides similar to the most popular ones already in use.

From idea to reality

ESG Companies contacted Tectonic Systems, Inc. President Ed Critz to research the market for applicable data collection systems. He selected Applied Vertical Systems, Inc. to supply the software, a time-and-attendance program and Theme Park Automated Management Software (TPAMS).

Computer Identics bar code printers, Printstar 60's were chosen for point-of-sale generation of bar-coded tickets with Code 39 symbology. In addition, Computer Identics readers, SCANSTAR 211's, are used at the rides to scan the tickets and collect the data via the Computer Identics STARNODE Network connecting the readers to the PC STARNODE board. Novell Network runs on an IBM PC local area network system. Eureka Systems RF transmitters in the Grand Prix track and RF sensors in the Grand Prix cars round out the system at MotorWorld.

Transparent but valuable

The automation at MotorWorld is nearly transparent to visitors, but the gathered data is a major management tool.

"It has been extremely helpful in the time and attendance and scheduling. It eliminates your typical time clock system where you have someone adding up the hours at the end of the week," says Andrea Kilmer, CPA, treasurer and comptroller of ESG. "The second thing that has been one of the most useful features is the accumulation of data concerning the type of tickets and combination of tickets, coupons and discounts received at the ticket window. It has allowed us to analyze our advertising and promotion projects form a monetary standpoint. That would have been impossible without the computerization," she says. "This (1988) is the first year since 1985 that we could monetarily evaluate our advertising."

"As a management tool, we love it," says Kilmer. "We have a very small management team. By having data so readily available, it helps to keep the staff down—to make decisions without an insurmountable support staff—and keeps overhead under control.

At the ticket window

When tickets are purchased at MotorWorld, the ticket sales person is prompted through TPAMS point-of-sale system to issue the proper number of tickets and for which ride(s). The terminal gives the total cost. If a discount coupon is submitted, its bar code is scanned. The computer verifies if it is valid for that ride on that day and deducts the discount. The cashier enters the amount of money given her by the customer and the screen displays the change.

Bar-coded tickets are printed in sequential order throughout the park so booth A may give number 001 and booth B may give number 002. While the ticket is printed, the guest is queried for his zip code and it is key entered into the terminal. The system ties it to the ride ticket number and coupon number along with the date, time, day of week and ride cost.

As the customer uses the bar-coded ticket at a ride, the ticket is scanned by a stationary reader. The bar code, date and time is sent, via the STARNODE Network, to the PC. That ride number is then killed, preventing someone from using a used, discarded ticket. The software then associates the purchase date and time, zip code and ride with the ticket bar code number.

The data is used to reconcile cash drawers and to determine tickets purchased and peak time for purchasing tickets. MotorWorld's management can analyze the results of promotion and advertising campaigns by zip code and use the data to create future campaigns. For example, coupons for a specific ride can be offered on beverage bottles or at fast food outlets based upon the ride's popularity in that zip code. In addition, peak ride-use times are used for employee scheduling.

Zoom with RF

The Grand Prix ride is where state-licensed drivers can perfect their skills driving Grand Prix-modeled cars. The driver must purchase a seasonal or lifetime MotorWorld Grand Prix Driver's license with the driver's picture and Code 39 bar code. The sequentially numbered bar codes reference the holders name, age, address, number of times he or she's been on the Grand Prix, lap times, average lap time and cumulative lap time.

Whenever the driver wants to drive on the Grand Prix track, a ride pass is issued, the license bar code ID is scanned and the car's ID number is keyed into a SCANSTAR 310 bar code terminal. When the car is driven over the starting RF transmitter, an RF signal is received by the software and the start time is recorded. When the car is driven over the ending RF transmitter, the end time is recorded. When the driver leaves the car, the license bar code is again scanned.

The software has collected the transmitted information and now produces a written invoice for the driver and the driver's name and address, date, lap times and amount due—with any submitted coupon discount subtracted. The invoice becomes a souvenir and the stored data is used by management for ride-use analysis and cash drawer reconciliation.

Race off winners

The data also finds a promotional use. The quick RF transmission of the data and the analysis of the data by the software was both obvious and necessary during Virginia Beach's Neptune Festival held in November 1988. MotorWorld took the information gathered over the summer and found the top speeds for the Grand Prix track.

The top racers, whose names and addresses were garnered from the driver license data, then were invited to a race off, with the final race held as part of Virginia Beach's Neptune Festival. Prizes were offered top racers in several categories. As in a normal race, the driver's bar code license was scanned and the race car's ID was entered. The lap start and stop times were sent by RF to the mainframe, a General Automation Zebra 5820. The software analyzed the data, and within seconds of the completion of each race driver's laps, the standings were updated and reported. The awards were presented at a ceremony attended by the Neptune Festival king and his court.

In addition to the sophisticated ride and POS revenue data collection at MotorWorld, the time and attendance program gathers staff work attendance. Employees are issued bar-coded, picture ID badges. Employees scan the badges at a SCANSTAR 211 reader located in the manager's office at the start of their work day. The badges are printed on standard card stock by Tectonic Systems using a Canon laser printer with a faster master controller. Captured data is used to manage attendance trends and for payroll purposes.

Slowing down speeders

Garcia plans to expand the automatic identification uses at MotorWorld. As bigger and more sophisticated race tracks are designed, Garcia and Applied Vertical Systems are working on RF data collection for analysis of speed and track conditions, with responses to include decreasing speed of cars when conditions demand it—such as a reckless driver.

In the future, it may also be used to locate lost or separated people or groups. If a group member has some of the tickets shared with the missing person, a message can be tagged to the adjacent numbers to alert cashiers the holder is missing or sought by a fellow group member. When the ticket is turned in to the cashier and scanned, the display screen flashes the appropriate message. A phone in every ticket booth allows the cashier to obtain further instructions.

Successful use of automatic identification in MotorWorld has spurred ESG to institute it in new theme parks and activities at Ocean Breeze. When a 36-hole mini golf course and a batting game open this spring, Auto. ID will play a major management and operation role.

MotorWorld's automation give it an edge in marketing analysis, up-to-the-moment ride use and customer traffic. Further, management can report on staff attendance and compare peak customer traffic to staffing requirements. Advertising results are quickly available.

As Ocean Breeze grows in size, success and sophistication, so will its use of automatic identification. It proves that Auto. ID is valuable for both work and play.

Comment: 4.8

Although the next article involves tags attached to automobiles, it should be categorized as process control, or more specifically, as work-in-process. The tag is only attached to the vehicle while it is being built, and is removed for reuse when manufacturing is complete.

Note that the term decoder is used in place of reader.

4.8 AUTO ODYSSEY

RF Tags Track Cars Through Baths, Paint, and Bread Pudding

Rip Keller—ID Systems—© March, 1987

Not long ago the flurry of paperwork needed to keep an automobile moving along the production line was a manager's nightmare. Though robots had replaced direct labor over much of the factory floor, the amount of indirect labor needed to coordinate the robotic operations remained vexing. Smooth functioning of the production line often required that a piece of paper saying "Paint this car red" arrive on the line as much as a week before the car itself showed up at the paint booth.

The system was not only labor intensive, it was error-prone, opaque to management who wanted to know when that batch of red cars was going to be ready, and not easily responsive to sudden changes in demand. All of that has changed at a new Austin Rover plant in Oxfordshire, England, where thermally insulated radio frequency (RF) transponders manufactured by Eureka Systems, Inc. are being used to track future automobiles through the production process.

Why use transponders? "Because when we used punch cards they used to get lost," says Steve Curtis, product support manager at Istel Ltd., the Rover subsidiary where the system was developed. Why not bar codes? "Bar codes are notorious for not liking spray paint." How about multi-layered bar codes that can be peeled and re-peeled so that they remain pristine in the grubbiest environments? "You can only use them once. Our bottom line doesn't approve of such profligacy. The transponder system is efficient, economical, and flexible."

The journey begins when the transponder is hung inside the right front wheel-well of a rudimentary body entering the on-track area. The worker performing the simple operation punches a key to inform the plant's DEC PDP 11/84 computer as to the car's basic body-type. There are four basic models in the Rover "800" passenger car series: the simple "GL," the fuel injected "I," the high-spec electric-windowed "SI," and the top-of-the-line "Sterling."

The body now passes a decoder. Activated by the decoder's electrical field, the tag wirelessly announces itself. Though the PDP 11/84 could poll the plant's decoders, Rover has opted for another method of getting the information to the computer, in order to save computer time. A Eureka Network Interface Unit (NIU) polls the decoders, regularly interrupting the 11/84 only when there is data to enter. Thus, when the NIU learns from decoder #1 that a car has entered the on-track area, it passes that information on to the computer.

The 11/84 has already predestined the car's future: it is to be automatic—extra brackets needed—or standard; it is to be painted in one of 16 colors or in a two-tone combination; it is to have left-side or right-side drive; and so on. The 11/84 will ride herd on the car all the way down the track, making sure at each station that this destiny—ultimately pre-ordained by management's view of the market—is being fulfilled.

There are, all told, 17 decoders on the journey. Nearly half of these are devoted to the rigors of painting, which, more than anything, mandated the transponder system.

Painting begins with an electroplate primer. Hanging from an overhead sling, the body is dipped into a high-temperature solution through which an electric current passes. The car is removed from the bath and air dried, then baked at 160°C.

Don Elliott, engineering manager at Eureka Systems, explains that hostile environments are not new to the "311" tag that Rover chose for this purpose. The tags were designed for such conditions as burial under concrete highways, where soaking, vibration, and frost heaves are among the tribulations encountered. High temperature, however, was not in the initial plan. So the tag, which measures approximately 1 by 1 by 1/4 inch, was encased in a cylindrical jacket measuring 3 1/2 inches in diameter and 3 1/2 inches in length with a connecting tab of 1 by 3/4 inches. The units are welded together ultrasonically and contain various oxides of high insulation value.

Thus encased, the long-life lithium battery that powers the transponder can withstand up to 45 minutes of 200°C conditions. This is not to say that the batteries last as long as they would at normal room temperatures, where they would be expected to provide a million reads before wearing out: Rover counts on no more than a few hundred production cycles from each tag. In the course of a normal work day the tags reach internal temperatures of 80 to 90°C. But Curtis explains that even at 100 production cycles, the transponder beats out bar code for the bottom-line prize.

Following the electroplating ordeal, car and transponder head for a coat of sprayed primer. Quality control requires inspection after each stage. Hence the car is identified upon entering the paint area, after electroplating—where it is either approved and sent on, or taken out for reworking—after the subsequent bake, after the sprayed primer, and again after baking. Now, if the car has passed quality control at each stage, it enters the color-painting area. If it failed, the computer is aware of the fact and will track it as it is reworked.

In the color shop, another type of problem emerges: cars must be scheduled in batches of the same color, and the batches must be as large as possible, because switching the spray from one color to another requires inordinate time. Unless the airborne paint residue from the spraying operation is allowed to settle thoroughly, it will color-contaminate following batches. This means that ideally all of the week's red cars—with their multitude of different options and features—get to the spray booth at the same time. And that is where the transponder system, with its reliable and instantaneous link with the host computer, becomes so important for overall plant efficiency. Because the whereabouts of each future car are reliably known to the computer, cars can be set aside in storage areas and brought back on line in color batches without disrupting the movement on the production line.

A corollary is that when paint-baking ovens break down—not a terribly frequent occurrence, but nonetheless economically significant— disruption to the production process can be minimized by the computer's capacity to reschedule production events effectively on the spur of the moment. When cars are shunted into off-track storage areas for these purposes, their whereabouts upon exit and re-entry are, of course, tracked once again by the computer via decoders at the entry and exit ports.

There is one problem in the paint area of the Rover plant that has not been solved and is apparently unlikely to be solved. Due to a peculiarity of the plant's organization, when the lunch-time siren sounds in Cowley, Oxfordshire, England, cars that are baking in the paint ovens simply remain there, while the workers hurry off to bread pudding and other important things. The transponders, in the meantime, get hot under the collar, and the stress reduces their life expectancy. A small price to pay, perhaps, for a humane workplace, but Eureka's Don Elliott does wince ever so slightly when the subject comes up.

Once complete and painted, the body goes to join its other parts— engine, suspension, fascia, rear window, wheels, and trim—on the assembly line. Rover uses the Eureka 311 tag to track pallets of some of these other parts as well. After all have become one, the system initiates the christening of the automobile, embossing the vehicle identification number (VIN) plate, stamping the number onto the body, and finally, printing two bar code labels for inventory purposes. The car passes off the track, and a final decoder notifies the computer that the car has been built. The PDP 11/84 passes the good news to the plant's PDP 11/73, which is programmed to control inventory and orders.

The system is one of the most advanced in Europe and has performed well for its happy owners. Don Elliott remembers but one bug in the initial installation: arc-welding equipment was generating noise that interfered with the transponder. This contingency had been overlooked in the planning, but placing the electronics differently eliminated the bug painlessly. So relax, harried managers—the paper chase has given way to the RF tag.

Rip Keller writes on communications, education and the arts for newspapers and magazines in New York and New England.

Comment: 4.9

The next article is similar to the Auto Odyssey article except that a completely different RF/ID technology, (SAW) is used and the concept of leaving the tag on the vehicles for multiple uses in the future is introduced. Also the temperature environment is more hostile and probably precludes the use of a battery.

4.9 RF TAGS

DRIVE CHRYSLER AUTOMOBILE PRODUCTION

Automatic I.D. News—© February, 1989

An RF/ID system at the Eagle Premiere plant helps create a top-quality, world-competitive vehicle.

The automobile assembly plant is a rugged, rugged place; not just any new technology can find a home there. No hair-trigger tempers or pampered pets need apply.

But RF fit the bill at one Canadian Chrysler plant, and in the bargain gave the company even more than it had hoped for.

When Chrysler Corporation's Bramalea, Ontario, plant went looking for a new identification technology that could withstand the rigors of its paint and trim departments, it found it had gotten even more—a system that provides manufacturing flexibility and more accurate ordering capability as well.

The story begins when the body and assembly plant selected an automatic radio frequency remote identification system to manufacture its Eagle Premiere. The RF identification system, from X-cyte, Inc. (XCI), provides accurate automatic vehicle identification (AVI) at critical process control points, allowing for real-time automatic control of the assembly process.

The Bramalea AVI system has vehicle identification points with XCI readers in the body assembly shop, the paint shop, and the final trim shop. As the vehicle assembly process begins, the Vehicle Identification Number (VIN) is written to the XCI transponder which is secured to the underside of the vehicle, on the radiator cross member frame.

At each vehicle identification point, an XCI reader mounted in the factory floor identifies the specific vehicle and automatically signals process controllers to engage the proper sequence of operations. Body parts, paint colors, trim parts, and so on are automatically matched. What's more, new parts are ordered automatically from inventory or from the parts suppliers.

The Chrysler process control system in place at Bramalea uses IBM PC/AT's and Tandem software to communicate with the IBM 4381 that serves the plant process control architecture. This computer-integrated manufacturing system, and the XCI AVI system, give Chrysler manufacturing fully automatic on-line process control capability.

Quality improvement crucial

"Manufacturing efficiency was our major consideration in moving to an automatic vehicle identification system," says Glen LaForet, production control manager at Bramalea. "With its excellent performance at Bramalea, the radio frequency AVI system has helped make Chrysler world competitive. The quality of our finished vehicles is significantly better. The instances of mismatched parts and need to rework have been reduced dramatically. And, we have made these accomplishments while reducing costs. This capability is part of why we think Bramalea is the most modern automotive assembly plant in North America."

Agrees Gerry Straughen, corporate C.I.M. system manager, "We saved the equivalent of about four cars a year by reducing reworking."

LaForet adds, "XCI had provided excellent products and thoroughly competent technical support, which are critical ingredients to the success of an AVI system. XCI people want customer satisfaction, and it shows. This is definitely a more efficient system."

Harsh environment test

Critical to the choice of the particular RF/ID system chosen was its required read range—"a foot is all the read range we really need," says LaForet—as well as read accuracy, overall system flexibility and simplicity, and cost effectiveness.

But a pressing need was the ability of the transponder to withstand the harsh environment of the paint and paint cure process. Initially, for instance, Chrysler had tried bar code in the vehicle assembly process, but with only limited success. Bar code did not adequately survive the harsh environment, particularly in the paint shop. RF/ID overcame the inherent disadvantages of bar code for this grueling environment.

The XCI system does this with its Surface Acoustic Wave (SAW) transponder technology. The reusable transponders have already survived more than 1000 cycles through the rigorous process at Bramalea.—The recyclable transponders, mounted with easy-to-use fishbone tags, are applied at the framing points, and taken off after the car is built, with encoding of the data for each unit done off-line.— Each of the cycles these transponders take, exposes them to alkali and other chemicals, to electrocoat baths, to paint, and to oven temperatures in excess of 400°F.

Labor savings

But in addition to withstanding this grueling reuse, the transponders also were found to reduce personnel expenditure while increasing accuracy. "There are many areas in an assembly plant where, without an AVI system, a person must sit and enter numbers all day," says Straughen. "This can involve more than five people per shift at each plant. When we built this plant, the RF system saved us from having to hire 15 new people."

"Plus," says LaForet, "the transponders are far more accurate and flexible. Before, there were only four or five sites where we recorded identification information. We never knew the particular situation between these points. But with transponders we have all the read points we need. This enable us to locate the vehicles at all times, and to know more accurately what equipment is required to complete them and to have it delivered in time to support the process without maintaining additional stock."

System flexibility is also key. LaForet notes, "RF also give us the potential to resequence the manufacturing process in certain locations to better suit our manpower or material requirements. Based on the information the system gives use, we can prepare schedules in advance and space our people more efficiently, maximizing manpower and enhancing Chrysler's commitment to quality."

This quality is assured as well through the automaker's corporate Performance Feedback System (PFS). At Bramalea, the PRF system involves RF. The Ontario plant key enters quality control information into the overall system, with RF being used to assure that the vehicle identification is accurate. Explains Straughen, "Chrysler uses AVI at Bramalea as input to our corporate Performance Feedback System, which monitors quality. We are now preparing to use electronic AVI at several new plants."

Other uses of RF are also being explored. In addition to scheduling and tracking vehicles, Chrysler is continuing to expand its ability to control the process in real time, thus eliminating much of the remaining paperwork associated with the building of vehicles. It is also considering the possibility of using an assembly process transponder which can also be left on the vehicle for shipping, distribution, and vehicle servicing.

"While other auto manufacturers also use the XCI system, Chrysler Bramalea had more fully utilized the technology to achieve complete automation capability." says Dennis Murray, vice president, sales, for XCI. "We're pleased that our system and our people are a part of this program at Bramalea. The system is working extremely well. And the new generation XCI systems introduced since the installation at Bramalea will make our contribution to the next Chrysler facility even better.

Comment: 4.10

> *Information written into a tag at multiple points can, and does, become a component of the total manufacturing data base.*

4.10 GM TURNS TO RF/ID FOR Q.C. AND W-I-P

Totally automated assembly line presents challenge

Automatic I.D. News—© February, 1989

Maintaining accurate and timely work-in-process (WIP) data for quality control in a totally automated, 2.8-liter aluminum engine head assembly line was a real challenge for General Motors' Chevrolet-Pontiac-Canada Group Engine plant in Buffalo, NY—especially for an assembly line that handles over 200 parts per hour to produce 5,000 heads each day. To address this problem, GM chose a radio frequency identification (RF/ID) system to collect WIP data.

Real-time WIP information

The RF/ID system, purchased from NDC Automation, uses totally passive reprogrammable tags, each carrying up to 64 bytes of information. The tags are integrated into a management information system (MIS) that tracks in real time the status of individual heads from start to finish along the 25-station assembly line.

Engineers at the plant are satisfied with the tracking system's results. The radio frequency tags and MIS combination keeps real-time, detailed WIP information, says Henry Voelk, assistant superintendent of manufacturing engineering.

"This is a totally new assembly line and a completely different concept of conveying information for us," says Voelk. "The old assembly line was more manual. The concept behind this system is that no human hands need handle the product."

Previously, GM used mechanical and electrical devices on the pallets to track production. A flag on top of the pallet was automatically set by an air cylinder as it passed through a workstation. The flag indicated whether or not work had been performed on the head. However, many times the flags would stick, wear out or simply not be activated, giving incorrect information. And since inspections were conducted manually at the end of the line, human error also was a problem.

Now, the RF/ID system is one of several components that ensures each head coming off the assembly line has all its parts and each part is correctly attached. In addition, inspections are now conducted throughout the assembly line by robots or automatic gauging stations, eliminating the need for human inspection of each head at the end of the assembly line.

RF/ID streamlines manufacturing process

Unassembled aluminum engine heads travel on pallets along the GM assembly line. An RF/ID tag, measuring approximately 2 by 4 by 5/8 inches, is attached to the bottom of each pallet. In GM's application, only two conditions, or "flags," on the tag are used: "pass/fail" and "work performed." They are stored in the tags for each of the 25 operations performed along the assembly line.

"We've found that the simpler you keep it, the better," explains Barry Goodin, supervisor, manufacturing engineering. "Since every head on the line is the same and is treated the same throughout assembly, we don't need to know when it was made or its size, color or pallet number. All we need to know is whether the work was performed and if it was performed properly."

Reader/writers—transceivers—located under the 25 workstations and at each inspection point, "read" the tags, sending the data to the main process controllers, supplied by Modicon Inc., a unit of AEG Aktiengesellschaft. When a head enters a workstation area, robots perform their particular assembly function. The transceiver then "writes" to the radio frequency tag that work has been performed for that station and the pallet proceeds to the next workstation.

To ensure product quality, four inspection or "decision-making" points are located throughout the assembly line. Here, inspection robots verify and test work previously performed on the head at several workstations.

If the head passes inspection, the decision-making point transceiver activates the tag's pass flag and it continues through the assembly line to the next workstation.

If the head fails inspection, the transceiver activates the tag's fail flag(s) for the faulty work procedure(s) and immediately diverts it to a repair loop.

Here, an operator inspects and corrects any flaws. Since one inspection robot verifies several work procedures, a repair loop transceiver transmits information stored in the tag to an inspection station information panel. The operator can then determine exactly which work procedure(s) was not performed properly. Before leaving the repair loop, the transceiver rewrites to the tag to say the work has now been successfully performed; and the pallet reenters the assembly line.

At the end of the assembly line, the fully-assembled heads are unloaded and transported overhead to the next 2.8 liter engine assembly area.

Small tags, enormous benefits

For GM, the benefits of the RF/ID system were quickly realized. "We now have much better control of the product throughout the entire assembly procedure," says Voelk. "The accuracy and reliability of the radio frequency tags has been outstanding."

"We have some extra error checking built into the radio frequency system, like using odd parity for data transfer, and an address echo scheme that ensures error-free addressing," Goodin adds. "So, once we read or write to a tag, we're 100% assured it's error-free."

"The distributed memory concept works very well for us. We have 12 controllers on the floor," says Goodin, "The RF/ID system relieves a great deal of the overhead for our main process controllers. For instance, once we write to a tag, we don't have to keep any of that information in the main process controllers. All of that information is stored in the tags until we need it."

The system cannot be fooled by humans because a work procedure is flagged immediately after it is performed, and each decision-making point checks all previous workstation flags. For instance, if someone removes a pallet from the assembly line, moves it past a workstation and then places it back in the assembly line, it will be caught at the next decision-making point.

Since the tags are noncontact devices, have no moving parts, and are read by units on the shop floor, they also fit well with other equipment, such as the asynchronous chain conveyor drive that moves the pallets and 28 robots used to assemble and inspect the heads. Tags also are immune to any electrical or magnetic disturbances and are unaffected by dirt, grease, water and extreme temperatures.

"The tags give us flexibility if we were to ever change our assembly line process," adds Voelk. "It would simply involve reprogramming the tags."

At GM, radio frequency technology has more than met the challenge of successfully maintaining accurate and timely WIP data for quality control in its totally-automated assembly line.

According to Voelk, the ultimate goal of the new aluminum-head manufacturing line is to produce a zero-defect product. In the two years it has been in operation, not one aluminum head has come off the assembly line with a defect resulting from the RF/ID system.

4.11 APPLICATIONS OF RF DATA STORAGE TECHNOLOGIES

The Need for Encoding Work-in-Process

Michael J. Hilligas—I.D. Expo—© 1988

Anyone who has stored or moved production data inside a programmable controller or computer can tell you that it can be one of the worst experiences of their lives. Lost memory or mistriggered hardware, which can shift the data, can become a nightmare. There was a time when a single bit could define the status of a part, and there are still a few machines that can be designed this way, but not many. The implementation of more and more complex automation, with ever increasing levels of quality control and diagnostics, has generated a great need for carrying the production data on the part carrier. This data requirement has grown in the past few years. I will in the next few pages describe some of the many ways that the different types of RF equipment can solve your application problems, and provide the level of performance needed.

Read Only Equipment

To eliminate one of the problems of data storage, which is the mistriggering of hardware, a READ ONLY system can be applied. This type of system can reference a carrier number, which is hardwire coded onto the tag at the time of assembly, with an internal file that will contain the data required for processing the product. The data storage requirements are quite high due to the fact that all data must reside in the main system, and system software must be as bullet proof as possible to ensure the data is not corrupted.

This type of system can also be used for location sensing on equipment such as Rack Storage & Retrieval Systems and Automatic Guided Vehicles. The tag is encoded with the location or bin number, read by the vehicle, and used to generate an action determined by this sequence. Some specialized products are available that will, when read, give you, in addition to the tag data, 3 digital lines for accurate position sensing. These signals when decoded can give you accuracies to within a few millimeters. This system does not account for the loss of internal memory due to tampering or failure.

Read/Write Equipment

To decide to use read/write equipment can completely change the concept of data storage within the production environment. The data capacity of this generation of RF Tags can be as small as 64 bytes to the current maximum of 32K bytes. These capacities can redefine the file organization of your system. It is no longer necessary to keep maximum data availability at any location where an antenna is mounted.

This type of system can also be used on equipment such as Rack Storage & Retrieval Systems and Automatic Guided Vehicles. The tag is encoded with the location or bin number, the product code for that particular location, and any inventory information such as quantity and date marking. This will allow the main system to verify its data base to the tag data read by the vehicle. The vehicle can update the tag with the proper inventory counts, dates, and all product codes describing the part or parts stored at that location.

I have used these systems on many occasions to solve my data transport problems. Here are some of the applications for which I have used Read Only and Read/Write Technology. All of these systems are running in production using a variety of different vendors equipment.

1) Buick Engine Plant, Flint, MI

        A) Cylinder Head Transport   (Read/Write)

        B) Engine Block transport   (Read/Write)

        C) V6 Engine Shipping System   (Read/Write)

2) Corvette Assembly Plant, Bowling Green, KY

        A) Uniframe Part Kit Transport   (Read/Write)

3) G.M. Truck Plant, Oshawa, Ontario

        A) Axle Assembly and Transport   (Read/Write)

        B) Fender Assembly and Transport   (Read/Write)

        C) Sheet Metal Transport   (Read/Write)

        D) Doors Assembly and Transport   (Read Write)

        E) Instrument Panel Assy. & Trans. (Read/Write)

        F) Cab / Box to Paint Transport   (Read Only)

4) Hydramatic Plant, Ypsilanti, MI

        A) Transmission Assy. and Trans.   (Read/Write)

5) G.M. Car Assembly Plant, Doraville, GA

        A) L.H. Wheelhouse Transport   (Read Only)

        B) R.H. Wheelhouse Transport   (Read Only)

        C) L.H. / R.H. Side Aperture Trans. (Read/Write)

The above systems utilize the capabilities of the equipment in much the same way.

The data written to the tag was as follows:

1) Carrier Number   (2 bytes)

2) Production status   (2 bytes)
         A) Fully Loaded
         B) Partially Loaded
         C) Empty
         D) Finished Part
         E) Reject Part

3) Production data   (250 bytes)

Figure 4-1. Typical Station Configuration

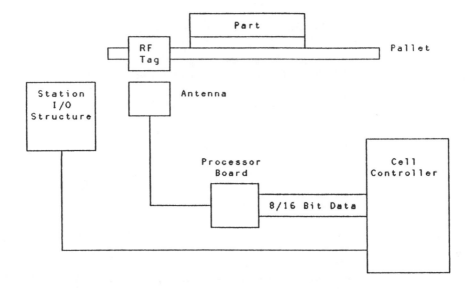

Data Configuration Techniques

The methods used in structuring the data on the tag can greatly effect (sic) the performance of the overall system. It is important that the words to be read the most frequently be put in consecutive order. Excessive cycle time may occur if too many reads or writes have to occur within one cycle.

A status word should be used to hold the "go, no-go" condition of the part. The next word or words can hold the part type. These words can be read at high speeds so that routing, or functions which are performed on the fly, can easily be accomplished.

Carrier Number: Initialize Value = 00h (hexadecimal)

The carrier or pallet number is written to this word. The function of this word is basically for maintenance logging. The main system can reference the number to a maintenance file for the so called Preventative Maintenance Program if any. This can be used for logging tooling defects, electrical and mechanical defects. The operator can enter the carrier number and tell the system to route the carrier to the repair area, and schedule a maintenance person for the repair.

Status Word: Initialize Value = FFh

Value should be set to an all reject state, and then set to the accept state upon proper completion of the intended process. Each bit within the status word corresponds to a process function.

Part Type: Initialize Value = 00h

Value can be an indexing code of part types for determining the direction of flow, or information transfer for process control.

Production Data: Initialize Values = 00h

This area of the tag memory is utilized for the storage of Production related data that would either be needed by a station to run a certain task, and to store the results of the task the station has just completed. The data can be test parameters such as: limits, percentages, program name, sequence number, serial number, option list, destination and so on and so on.

The production data section can be accessed with relative speed depending on the amount of data required for the operation. Typically the operations that need large amounts of data usually require the stopping and or the accurate positioning of the carrier or pallet to complete its operation. This will allow the access to as much data as needed.

Typical Tag Configuration

Carrier or Pallet Number—1-2 Words
Status Words—1-10 Words
Part ID—-2 Words
Production data—1-Maximum Words

As you can see there is an almost infinite possibility to utilize the memory, and with memory capacities as they are, the only restriction to data size is the cycle time of the system.

Future Memory System Technology

As the data storage requirements become greater, and more and more production information is being carried with the product throughout the production facility, the data required for a station function could become noncontiguous, and the need for multiple access to the tag will become necessary, slowing the system throughout.

I have talked to some of the manufacturers about the development of an intelligent tag. This tag would have the ability to be user programmed for internal data manipulation. The program would be able to, in much the same way that a spreadsheet is configured, alter memory locations using a defined formula.

With the program running, a one word function code can be set up for each read point. With this one word the reading station can perform all the necessary functions. This would enable the possibility of a much quicker cycle time due to the near instantaneous access to the tag. The station logic can now be freed from decoding a long string of data and perform a predetermined function referenced by the word read by the system.

This type of system is not currently available, but this is a fast moving industry, and if a need is shown for this type of technology the industry will respond.

I am currently developing systems using the stand alone technology. These devices have onboard a maximum of 16 inputs and 16 outputs, auxiliary serial ports for printers and displays, an antenna for tag access, and the ability to be networked together for host download of information or programs, and data transfer from the station.

These systems can aid in the decentralization of control in the production environment. The simplification of wiring and software, and the overall throughput increase due to local processing of data, can make this type of concept very appealing. All the station functions can be controlled by the reader whether it be solenoids, or interlocks with robots and auxiliary equipment, to the communications with CRT displays, printers, bar code printers or readers, and any other serial device.

Care should be taken in selecting a unit because some have the ability to format text easily, if a lot of ASCII transmission is to occur. Some units will run faster and have watchdog functions for better machine control. The ability to solve Boolean equations should also be a consideration.

I have used this system on the following machine.

1) Rockwell International Corp., Uniontown, PA
        A) Water Meter Test System

The Project utilizes the readers for a variety of applications. Here are some of the usages:

1) Load Station
        A) Scan Pallet Status
        B) Update Production Data
        C) Interlock with PLC through Digital I/O
        D) Serial Link to VME Host System

2) Test Stand (24 Inline Parts with 2 Readers)
        A) Read Test Parameters on 24 Antenna using its
            digital I/O to access separate tags.
        B) Update Production Data on 24 tags
        C) Interlock with PLC through Digital I/O
        D) Serial Link to VME Host System

3) Label Station
   A) Read Pallet Status and Production Data for Label
   B) Send Serial Data to Label Printer
   C) Interlock to Label Applicator
   D) Interlock with PLC through Digital I/O
   E) Serial Link to VME Host System

4) Unload Station
   A) Read Production Data
   B) Interlock with PLC through Digital I/O
   C) Serial Link to VME Host System

5) Reject Station
   A) Read Production data
   B) Interlock with PLC through Digital I/O
   C) Serial Link to VME Host System

Typical Station Configuration

The units can also be mounted on a moving vehicle such as an AGV or Storage System. They can both read the data and control the motion of the vehicle.

These devices have reduced my overhead in data communications to the reader system and hardware requirements. The throughput of the system has also benefited from the increased process speed. I think this equipment will be selected for more and more applications needing high speed response to tag data and local I/O control.

Installation Techniques

Although the installation of RF equipment is quite easy, care must be taken when the device is selected. There can be interference between antennas placed so close together that they will cause errors or functional failure of the device. Also the placing of 2 tags too close together can cause misread data or data corruption onboard the tag.

This can be seen in a system which has, for instance, a very close station to station spacing where antennas will have to be placed in close proximity with each other. Antennas that can have controllable field strength can solve this problem. The system software can also cause only one antenna to be active at one time within the field area.

There is a more serious problem, and that is having more than one tag within the transmission field. This can only be solved by lowering the RF transmission field intensity by using different antennas, or reducing the sensitivity of the tag. The problem may also be solved by mounting equipment at angles that are skewed from their optimal position. Although I do not recommend this approach, it can be made to work when faced with an impossible situation.

The quantity of reader/writers on one serial data link must also be considered. If the quantity of data to be transmitted is large, by that I mean 100 words or more, and the frequency of reading is high the number of readers on the link may have to be reduced. If your cycle rate is low, or the quantity of words to be transmitted is low, the data link can support as may as 8 to 12. I am used to very high cycle rate machines, 350 pph rates have to be handled. This is why my counts might seem low. In your applications there are some fundamental calculations which should be performed in order to come up with the maximum number of antennas (or readers) which can be connected to one data link.

Conclusion

This technology has really matured in the last few years, and shows no sign of standing still. I think a great future is in store for the industry. I see the products on nearly all the jobs that we do, and many others that I see in the field.

With a little imagination there are many uses for the equipment, and I hope you have found this information helpful.

For additional information contact:

Michael J. Hilligas
Wilson Machine/Allen Translift Div.
400 Florence St.
Saginaw, Michigan 48602

The Future of RF/ID

The future of RF/ID and Communicating Electronic Documents will be determined by many complex and interrelated factors. The articles in this chapter identify many of these factors and in some cases attempt to analyze and understand these interrelationships. When you have read them, try to project yourself into the future and hypothesize how RF/ID will effect your life—personally, professionally or both.

Comment: 5.1

> *The future is a moving target. Here is how it looked to me in late 1987. The title was chosen by an editor. My suggestion for the title was, "RF/ID, The Prologue is Passed (or Past)."*

5.1 RF Prophecy

An Expert's Vision of RF/ID's Future

Ron Ames—ID Systems—© December 1987

Radio Frequency Identification (RF/ID) began about 1973, but it is still an industry measured in only tens of millions of dollars. Even though it has taken 15 years for RF/ID to reach this size—small compared to other ID technologies—I predict that within the next three to five years the industry will be measured in the hundreds of millions of dollars.

Why go out on a limb by making such a prediction, given that several industry factors point away from such rapid growth? First, let's look at those factors. According to Radio Frequency ID: Technology, Markets and Applications, an industry research report which I coauthored, the average annual sales per RF/ID manufacturer is less than two million dollars. There are about 40 companies that manufacture RF/ID equipment or design systems. In most of these companies' business plans, the same RF/ID applications mentioned 10 or 15 years ago are still being mentioned. All indicators support a prediction of slow growth in RF/ID.

In contrast, the bar code industry has grown dramatically during the same period. One might conclude that bar code has satisfied all of the market requirements for automatic identification and that "all others need not apply." That would be a big mistake, and the 40 plus companies that will make up this industry for the next few years agree.

Cost vs. Benefit

Investors behind these companies have asked, and will probably continue to ask, "Why will the market for this technology suddenly take off and grow dramatically for several years?" A couple of precedents, with an important point in common, answer this question.

The first precedent is the hearing aid business. You may remember seeing hearing aids that were about half the size of a brick. They preformed the basic function of amplifying sound, but they were heavy, awkward to wear, impossible to conceal, and had very little sophistication. And to know one's surprise, they sold because some people had to have an improvement in their ability to hear. When hearing aids became small enough to fit completely inside the ear, had automatic gain control, and didn't blow scalps off with high-pitched whistles, they began to sell in high volume. The message is that when they achieved the threshold of cost/benefit ratio that was necessary for the majority of the potential hearing-aid market to decide to buy the product, they began to sell in volume.

The second precedent is the personal computer business. Single-user computers had been around since John von Neuman's little group managed to get the first one to do something useful. The computers were not personable—more like, hostile—and they were not personal in the sense that they were not bought and paid for by a single individual. Personal computers, like hearing aids, had to achieve the cost/benefit ratio that was necessary for the majority of that potential market to decide to buy, before high-volume sales could be achieved. More complex factors were involved in achieving the acceptable cost/benefit ratio in personal computers than in the hearing aid precedent, but the premise is the same and it also holds for RF/ID.

Before answering the question I posed, let me identify some of the applications that have been repeatedly mentioned in so many business plans over the years and indicate where RF/ID stands to achieve the cost/benefit threshold of general market acceptance.

Applications That Count

The market for security and access control systems trades off three factors: cost, security, and convenience. In my opinion, cost has been the most important factor, convenience ranks second, and real security ranks last.—I'm prepared to defend my opinion to anyone who cares to debate the issue.—However, RF/ID now offers the greatest convenience of any credential-based approach, the highest security—difficulty of counterfeiting—and the technology to be cost competitive with any other current technology, with a comparable share of the market. Not coincidentally, this is the market segment in which RF/ID has been doing best and in which it will continue to do well in the foreseeable future.

Work-in-process systems either carry instructions about the work to be done in a manufacturing process or log the work which has been done—sometimes both are done in the same system. The market has demanded systems that do these functions with reliability, capacity, speed, and cost, usually in that order.

RF/ID has offered acceptable solutions in many cases; however, there has been room for across-the-board improvement. The technology is available to provide improvement in all these areas, particularly in capacity and cost. Products with these improvements are continuing to enter the market. Other factors have made this a slow growth segment. One of these is rather slow incorporation of communications networks for use on the factory floor. Another is the availability of application software to drive the system. Both of these are much more easily addressed with quality solutions today than was possible even a year or two ago.

The market for material handling, or automatic identification systems that identify and track material in the factory, has been well-served to a large extent by bar codes. This will continue to be true in cases where the environment is relatively clean and where read/write capability is not required. In other situations, RF/ID can provide the solution. The technology exists to produce thin, flexible, self-adhesive RF/ID labels at prices less than two dollars each or, in high volumes, at less than one dollar. Corshield corrugated cardboard boxes with RF/ID capability integrated in them, with an incremental cost of fifty cents per box, also appears to be feasible in large volumes. These, and other implementations of RF/ID technology, should facilitate penetration of this market.

The cargo and baggage-handling markets share the challenge of labeling an object that already exists and having it perform well in a tough environment.

Cargo is predominantly handled in 20 or 40-foot containers which need to be identified from a vehicle driving through a storage yard in any weather.

Baggage-handling and sortation systems must handle bags in any orientation, even lying on their tags, at ranges of up to three feet. A recent article stated that a bar code-based baggage-handling system in Chicago was achieving a whopping 60 percent read rate. No wonder the number of airline complaints is up several hundred percent in the last three years.—Recent reports indicate they are up to 98%.read rate.—A system which can't deliver a 99.9 percent or better read rate should not be installed. RF/ID certainly is capable of this of this degree of accuracy.

These applications and others, such as asset management, motor vehicle identification—including automobiles, trucks, trailers, and buses for parking and toll applications—and trains for engine, car, caboose, end-of-train, and train location are all feasible today or will be in the very near future. Animal identification for gene pool management, automated care and feeding, disease control, pet recovery, and import quarantine control are also feasible applications today. All of these applications have been mentioned in several business plans and still have significantly less than 50% market penetration. In most cases actual penetration is less than 10%.

A reasonable conclusion is that although the functionality and, in most cases, the added convenience make RF/ID attractive, it has not achieved the cost/benefit ratio that most buyers want to see before deciding to buy.

Product cost is still an issue in some of the markets I just mentioned. It is also an issue in several markets, that have not been commonly mentioned in business plans until more recently, but which either require RF/ID or could significantly benefit from the capabilities it offers.

The Answer

One issue stands out as being crucial to the success of the personal computer market explosion and in the much higher growth rate of bar codes vs. RF/ID in recent years. That issue is standards.

There, I've said it. It's public, and before certain manufacturers can stone me in the town square, the discussion will have started. This is a genie which, once released, never goes back in the bottle.

In all of the markets previously mentioned, the users would benefit if they could buy compatible products from multiple vendors—with the possible exception of work-in-process. And equally importantly, the manufacturers would also benefit. This industry needs to address this issue quickly, forthrightly, and constructively. And I believe we will. The RF/ID committee in AIM has already started contacting interested user groups and vendors.

So the answer to the question, "Why will the market for this technology suddenly take off and grow dramatically for several years?" is this:

The technology and most of the applications are well enough known in the industry to produce functionally acceptable solutions, and if appropriate standards are adopted and standard products are brought to market, the attendant cost/benefit ratios will result in high-growth rates for several years.

Each of several markets represents potentially larger dollar sales than the current total RF/ID market . And there is good reason to believe that more than one of these markets will be penetrated by the next generation of products. I have seen the technology necessary to achieve it.

Given this perspective, predicting an RF/ID industry measured in hundreds of millions of dollars doesn't seem reckless at all—only sensible.

Ron Ames, a member of AIM's RF committee, is an expert on RF technology and a consultant in the field.

Comment: 5.2

When I wrote this next article, I expected to be denounced, or possibly stoned publicly, by a few manufacturers who see themselves as closed system suppliers, who would be threatened by standardized open system products which would achieve lower cost through volume and would take their customers away. This has not happened. Rather, it turned out to be a whisper campaign and a few Molotov cocktails thrown from the shadows. When de facto standards begin to appear as a result of multi-sourced supply to some high volume applications, I still expect to see some tantrums.

The other thing I expected was that user communities would surface and start developing application standards. Although I am aware of a few groups doing this, only one has approached the RF/ID Committee of AIM for assistance thus far.

It is fair to say that I am disappointed by the inaccuracy of at least one of these predictions but that is the risk one takes as a fortune teller.

5.2 Radio Frequency Identification (RF/ID) Systems:

A Media Waiting for Standards

Ron Ames—ID Systems—© April, 1988

In my last article—RF Prophecy, I.D. Systems, December, 1987—I answered the question,"Why will the market for this technology suddenly take off and grow dramatically for several years?" by stating, "The technology and most of the applications are well enough known in the industry to produce functionally acceptable solutions, and if appropriate standards are adopted and standard products are brought to market, the attendant cost/benefit ratios will result in high growth rates for several years."

I believe that RF tags can and will be viewed by the marketplace as another portable data storage media. I freely admit that this will not be true in some applications, predominantly those in which the objects being identified remain under the control of the owner and in which a single database is accessible to all Readers or Reader/Writers in the system. These applications are called CLOSED SYSTEMS. In these cases the owner of the system may well say, "Why should I be concerned about standards, I buy all of the equipment from the same company?" There is a good reason which I will address later.

In other cases, where there are multiple owners involved—e.g. objects with RF tags attached are shipped among the owners of the reading systems—or the reading locations are geographically remote and do not have access to the same data base, these are called OPEN SYSTEMS.

A really strong case can be made for treating the tags in these environments as another physical form of portable storage media.

In fact, they may be performing much the same function as other media technologies such as punched cards, printed Bar Code and OCR labels, magnetic stripes or floppy disks. In large shipments, I have even received a roll of half inch, 9 Track ANSI Standard magnetic tape which included a manifest, as well as unpacking and assembly instructions. We printed it, and distributed sections to the appropriate departments so their jobs, relating to the shipment, could be done. It was easy, because we had a compatible—STANDARD—tape drive on our computer.

If interchangeability is of any value or importance in an application, and if the application is replicated a large number of times, standardization should be considered.

Many applications may appear distinctly different from each other, however in respect to their labeling or identification needs they may be very similar. So similar, that in many cases, a product designed to meet one set of needs would be almost ideal for some others. In fact, one design might address many sets of needs, so well, that if they were all added together they might represent a very large market.

One of the primary functions of marketing is to define products so they are highly attractive to multiple application areas. Often they are called "standard products" as opposed to those which are specially designed for each customer or application.

The customers of these standard products benefit from paying a smaller share of development costs which are spread over the total number of items produced. They also benefit from the cost savings that come from building large numbers of the identical item. This is why you should be concerned about standards even if you intend to buy all of your equipment from one manufacturer. You can share in the savings earned by the combined buying power of all of the customers of the product.

The manufacturers also benefit from the savings. And, by lowering prices, they can create additional new markets in which to share.

We have all benefited from the lower costs, larger variety of products and larger number of sales and service locations in such products as audio tape cassettes, VCR cassettes, floppy disks, CD ROM's, phones, and personal computers, due to economies of scale that were made possible by the adoption and implementation of standards.

Many of us do not know how these standards came to be, but we should be prepared to question our suppliers about their participation in the development and implementation of standards in products that we plan to use. We also should let them know what our future needs are, so the standards developed will reflect what users will want to buy.

If you are a manufacturer and you are saying to yourself, "I want be a part of the solution, what are some constructive things that I can do to speed up the process?"

I'm glad you asked because that reflects an attitude of open mindedness that is essential to finding WIN-WIN solutions to some challenging questions.

The first thing you can do is to review some of the good marketing work you have already done. Identify applications you have encountered in the past which involved OPEN SYSTEMS—They are in your "Lost Business File."

Separate them into groups having similar requirements, then take photocopies of the Standard RF/ID Chip Survey Form, Figure 5-1, and fill it out.

Sure, it's OK to have the Engineering guys help you fill it out. It's mostly their turf anyway. I guarantee that you will instantly be involved in the most stimulating, spirited and far-reaching discussions you have had since the CFO was called out on a close play sliding into third at the company picnic.

In the company history, it will be remembered as a milestone in company culture development, a meeting in which the company took a monumental step toward being a major player in what became a major industry.

While most of the blanks are self explanatory, the Tag Category needs some definitions. Tag Categories are made up of two attributes, Class—which refers to capacity—and IQ—which relates to whether a microprocessor is incorporated—Smart—or not—Dumb. The priority is your ranking of the market urgency of having this chip relative to the others you have identified a need for and described on other sheets. —I think six priority levels are probably enough.—

The Class definitions are as follows: Level II is 8-128 bits, Level III is 48-512 bits, Level IV is 256- to the maximum number of bits available.—I expect additional levels will be defined in the future.—

Let me address the implications of the last section, Ownership. One of the toughest challenges for which we have to develop a WIN-WIN solution, is the ownership of the technology represented by each of the TECHNOLOGY STANDARDS that will be developed from information collected on these forms and other sources.

By way of encouragement, this problem has been successfully solved by all of the other storage medias, data communications, peripherals, radio and television—both monaural and stereo—CB radio, batteries, as well as others too numerous to mention. So I am confident that a fair and amicable solution will be found.

When you have filled out a form for each tag category you believe is needed, send them to :

RF/ID Committee
AIM
1326 Freeport
Road,Pittsburgh, PA 15238.

Enclose a letter telling them you would like to participate in the development of meaningful standards to aid the growth of the RF/ID industry, and that you plan to attend the next committee meeting. This committee is dedicated to the health and growth of the industry and is taking a leadership role in the development of standards.

Another thing you can do is encourage the companies and trade organizations who have applications which need multisourced standard products, to contact AIM for information on how to they can develop APPLICATION STANDARDS.

If you are an actual or potential user of this technology and you think others in your industry also need it, work with them, through a trade organization, if there is one in your industry, to develop APPLICATION STANDARDS. These APPLICATION STANDARDS are another source of data to to help the manufacturers develop the TECHNOLOGY STANDARDS for the products you and other markets need and to justify the investment required to provide those products.

MARKET DRIVEN STANDARDS are worth our effort to develop, because whether we are the producer or consumer, we all benefit from them.

Figure 5-1

Standard RF/ID Chip Preference Survey

Name:_____ Phone:(___)_____ Date_____
Address:_____ St_____ Zip_____
Tag Category: Class; [ ] I,[ ] III,[ ] IV   IQ; [ ] Smart, [ ] Dumb Priority (1-6)_____

Frequency

Band Preferred: [ ] Very Low   [ ] Low   [ ] Medium   [ ] High
                      50-150 kHz   250-500 kHz   1.70-28.0 MHz   900 MHz+
Freq.To Chip: _____ Freq. From Chip: _____

Sensitivity
Minimum Signal Strength at Chip: _____ V _____ A

Selectivity
Maximum Frequency Deviation Tolerated by Chip: _____

Power
Min. Power to Operate Chip: _____ Power Source: _____

Modulation
Type Modulation to Chip: _____ From Chip: _____

Protocol
Description of Protocol: _____

Error Management
Description of Error Management: _____

Commands
List and Describe Commands the Chip Should Recognize: _____
_____

Sequences
List Valid Command and Data Sequences: _____
_____
_____

Ownership
To Your Knowledge, is the Technology Described Above Covered by a Patent or a Patent
Application?_____ Who is the Owner or Assignee?_____
_____
Other Comments: _____

Comment; 5.3

> *This article was written more than a year after the RF Prophecy article
> for an audience primarily interested in factory related applications. My
> suggested title was: "RF/ID, the Next Five Years.'*

5.3 Radio Frequency Identification (RFID)

in Manufacturing '89

Ron Ames—Material Handling Engineering—© January, 1989

My vision of Radio Frequency Identification's role in factory applications for the next five years is just radical enough, I believe, to make some managers review their own five-year plans to see whether RF/ID should be considered. If I can cause some managers to visualize RF/ID in their future, this article will have achieved its goal.

I also happen to believe that there will be significant advances in the use of RF/ID tags—defined as a device that uses some portion of the electromagnetic spectrum to communicate to a small remote data base, is an RF/ID system.

Those knowledgeable in the field will immediately realize that I have broadened the prevailing definition of RF/ID. The reason I have done so, is my awareness of the development of technology and products which fit the broader definition and are functionally equivalent, in most respects, to current RF/ID tags.

There are five categories of RF/ID use within the factory that I will specifically address: material handling; work-in-process; asset management/tool ID; job costing/labor scheduling; and product distribution and control. I have also broadened some of these definitions, as explained later.

Material handling and some of the other applications have been progressing well in their development and currently use bar codes for their Auto-ID needs. Why introduce the idea of using RF/ID at this stage?

Consider this statement: The economic justification for any portion of the manufacturing process must be the assessment of its contribution to the objectives of the entire process against any alternative means. An appropriate set of accounting policies must be consistent with this.

Apply this thinking to the Auto-ID requirement where the whole whole process of receiving, storing, retrieving, and tracking material and that of controlling the manufacturing process is complex, as well as exception prone. With respect to Auto. ID, the identity of the entity alone does not sufficiently define its status at a particular moment in time. Therefore valuable and sometimes crucial time loss can be avoided if a sufficient amount of historical, specification or other data can be accessed instantly where the entity in question is located.

In some applications, therefore, the real cost of a bar code system should include the cost of working around the fact that they are not easily changed or updated.

The apparent low cost also makes it too difficult to change from bar codes to a more "objective efficient" method of distribution, proactive, manufacturing process control.—Objective efficiency is a term I invented which is intended to reflect not necessarily pure cost, but the most efficient way of achieving whatever the selected objectives are, e.g. quality, availability, life cycle cost, manufactured cost, installed cost, etc.

Complicating the issue is that technological progress is rapidly closing the gap between the cost of bar codes and both read-only RF/ID tags and RF read/write tags. The need to look at the big picture, in order to select the right technology to provide the optimal capability, becomes greater.

Here's my forecast for the previously mentioned five areas of use.

Material handling.

Tags molded into the pallet, tote or container can last the life of the container so the amortized cost can be low. They can't fall off and they can be instantly updated as material is added or removed. Portable reader/writers can upload changes they have made to an on-line data base. I project some penetration of this application over the next 5 years but probably not more than 15 percent of the total market. Even that is a large dollar volume.

Work-in-Process

RF/ID is already having significant success, particularly in the automotive industry, and I expect this to continue at an increasing rate. The instructions for building the product, a record of progress thus far, and test data may all reside in the accompanying tag. In five years, RF/ID may have the majority of this application, not just in rust belt industry but in electronic equipment and some other fields as well.

Asset Management/Tool ID:

In many companies, tools are their major asset and it is where most of the software and systems analysis is currently being done. Tool ID has really just started; but almost immediately the emphasis has switched to asset management. The best use of the tool requires that machine tool adapters have the adapters identification as well as dimensional characteristics stored in them.

The identification prevents damage to the machine, tool adapter and the work piece by preventing the use of the wrong tool. Read/Write tags are used in the tool adaptors. Read-only tags will be used in components added to adapters and smaller tools.

I expect that in three to five years at least 20 to 30 percent of the older machine tools will have been retrofitted with this type of RF/ID system and that 70 to 80 percent of new machines will be equipped when shipped. Other assets will be tagged with identification, maintenance data and other relevant information.

CNC Program Storage

Another major application related to machine tools but which does not fall in any of the categories I have defined, is the media that is used to transport the CNC program itself. There is considerable resistance to down loading CNC programs on-line, and punched tape has some disadvantages. Floppy disks are also susceptible to contamination and damage.

The Japanese are pioneering the use of RF memory cards. These look like credit cards but hold almost as much information as a floppy. Since they are completely sealed, they provide an excellent solution to the problem. The Japanese project a multibillion dollar market for these cards. A fair number can be expected to reside in the machine shops of the world.

Job Costing/Labor Scheduling

Both deal primarily with the management of a resource called skilled labor hours. Labor scheduling has to do with the labor equivalent of JIT or having the right skill at the right place, with the right tools and material to perform particular tasks. This can be accomplished through the use of Read/Write RF Badges for the workers, which can have downloaded work schedule information.

This information can be updated as the worker completes tasks and as other variables affect the work flow. Job costing includes recording actual labor, not just standard labor, for elements of the work order. This is an excellent means of closing the loop on the human part of the equation and assuring that people are also at the right place at the right time and that they can do their jobs when they are there. Since system and software changes are required to use RF/ID, I expect fewer than 15 percent of these applications will be done by RF/ID, in the next five years, although it will be accelerating.

Distribution and Control.

As tag costs come down, more products will have them designed-in for permanent use. The so-called ship-with tag, which is manufactured into the product, will carry part of the work-in-process burden, recording options installed in order to accurately define the product as it will ship, for matching the order. It also is used to track its progress through, not only the distribution cycle, but through retail sales, the warranty period and the rest of its service history.

The tag cost probably can not represent more than 0.1 percent of the selling price, so you won't see them on potato chips. However, they will be used extensively on capital equipment and higher priced consumer goods within the next three to five years.

Tag Cost Forecast

What can the cost of a—not all—read only RF/ID tag(s) be in three to five years? As a manufacturer of RF/ID integrated circuit chips, the lowest cost I can project for a functioning but unpackaged read-only tag module, based on technology I have seen in the lab, is $0.15 to 0.25, based on silicon integrated circuits. A non-IC technology holds the promise of reaching a cost of less than a dime.

These projections require a high unit volume, to be sure, but we are getting close to seeing the necessary commitments in both volume and price. We are also at the point where the packaging of these modules may be the largest cost in a complete tag. If the tag module is injection molded into a pallet, tote or into the product itself, the packaging cost is much less significant.

What about a 1000 bit—128 character) read/write tag in three to five years? I believe they can reach $1.00 to $1.50 for an unpackaged tag module in that time frame. Larger storage capacities will track the cost of EEPROM's or ferroelectric memory chips.

Both of these tag module costs would cause strategic thinking manufacturing planners to seriously evaluate the incorporation of RF/ID technology in factory applications.

When to Start?

The question is: When should they start? The answer is that some already have started and have installed systems that are showing a good ROI. Others are in the evaluation process, sometimes fighting obsolete accounting policies, but still trying.

Still others are copping out by saying "When the tags get as cheap as bar codes, then maybe. Besides, I retire in three years, why should I rock the boat? I just hope the boss doesn't see that article."

Comment: 5.4

This article illustrates how new uses of technology become reality. This application has been stalled by lack of budget which will result in the incorporation of additional technological advances before it becomes generally used throughout the military.

5.4 "Smart" dog tags

may save lives on battlefield

Carl Kovac—Automatic I.D. News—© October, 1988

On some future battlefield, a badly wounded soldier's life hangs on a small plastic tag worn with his aluminum dog tags on a chain around his neck.

A combat medic crouching beside the semi-conscious rifleman inserts the tag into a hand-held reader. An LCD display gives him the GI's name, serial number and blood type. It also says the wounded man is extremely allergic to the pain killer the medic is about to give him. The syringe is quickly forgotten.

After staunching the bleeding, the medic uses the reader to enter the nature of the soldier's wound and the treatment he has just received into the microchip embedded in the tag.

Life saving data

A short, fast chopper ride later, personnel at a field hospital read this information as they prepare the soldier for surgery. The tag also tells them about inoculations he has received, any recent illnesses, past wounds and other allergies he may have.

What initially saved his life was an ICR—Individual Carried Record tag, developed under contract to the U.S. Army by Battelle Memorial Institute, Columbus, OH. Had he been wearing just dog tags—which give only name, serial number, blood type and religion—an injection might have accomplished what a chunk of enemy shrapnel failed to do.

Battelle, which has been involved in smart-card research and development for some time, began work on an electronically erasable and updateable ID tag to complement traditional metal dog tags about three years ago. "After the Army caught wind of smart-card technology, they asked us if we could come up with battlefield and peacetime applications," says Rich Rosen, group manager for electronic product development at Battelle. "They wanted something that would be no larger than ID card and no smaller than a dog tag."

Multi-use ID

Pertinent information would be added or deleted as circumstances required. The tag would enable soldiers to carry basic identification and medical information into combat. More information could be inserted to facilitate individual transfers, troop movements, access on and off military posts and myriad other uses.

Battelle initially developed a Soldier Data Tag (SDT) Design Guide. This focused on identifying tag materials that would stand up to the rigors of the battlefield; looked at potential security countermeasures for the system and their relation to personnel and national security; evaluated the compatibility of an SDT system with other Army information networks; and analyzed the cost-benefits of activating such a system in the modern Army.

Contact vs. non-contact

It then evaluated and compared potential commercial products to a prototype SDT. "We put credit card-shaped smart cards, optical storage cards, non-contact smart cards and the prototype SDT through electrical, chemical, mechanical and environmental stress tests," says Rosen. "Based on the results, we initially recommended that the Army consider non-contact technology. However, we were then asked to develop a prototype medical application for the ICR, using portable data carriers, so now the Army is looking at both contact and non-contact devices using integrated circuit and optical technology."

It will be some time yet before ICRs find their way around GIs' necks. Depending upon funding availability, the Army hopes to have an ICR system operational by fiscal 1990, Rosen says.

The Future of RF/ID

Comment; 5.5

> *This confusing article was intended to make one simple point. That, particularly in the field of electronics, functional attributes of products can easily be cut and pasted into previously unthought of combinations, if they result in something that is useful to someone.*
>
> *Therefore we should not be constrained to think about solutions to our problems only in terms or definitions that have become established or accepted in some official way.*

5.5 Confusion is sometimes a good state to be in

Ron Ames—Automatic I.D.News—© October, 1988

A few months ago I was asked, "What industry are we in?" Thinking about it confused me.

Over the past several years, as member of the Automatic Identification manufacturers (AIM) I have heard answers to that question that are unsatisfying to me, as well as, I'm sure, to others. There is a tendency to like definitions that include our area of expertise, especially if it makes our part look the most significant, most dynamic or the largest, even if it leaves out those with a claim to membership.

AIM brings together products with one common capability; they are used to identify things. AIM has been liberal in including products in the definition of automatic identification.

Most of the time Automatic Identification products can do other things, such as providing additional information about entities to which they are attached, or provide real-time communication links for a dialogue about the entities in question.

Sometimes, automatic identification may be a minor part of a products' complete role. Additional information would be of little value if we did not know to which entity the additional information applied, so identity is important to us.

From this common attribute—the ability to automatically identify—products of AIM members diverge. So, like others, I get confused.

Some companies define the industry as a data collection industry. This may be right if only a read-only media such as bar codes or optical character recognition (OCR) is included. If read/write media such as magnetic stripe, or RF/ID tags are involved, the description will not cover the product capabilities. I know it confuses me.

Differentiating products

One way to differentiate products is by the media of communication between labels or tags on the object and the devices that reads the data.

In some cases it is light—visible, infrared or ultraviolet. Examples of this are bar codes, OCR, Vericodes and vision systems. All use light, but some use ambient light, others a special wavelength light-source. Is this confusing?

Voice recognition and synthesis systems use sound as the media; in other cases, electromagnetism. This media is used in magnetic stripe, radio frequency identification (RF/ID) tags and RF smart cards.—While light and radio frequencies are part of the electromagnetic spectrum, their behaviors are extremely different. Significant behavioral differences appear between extremes of the radio frequency portion of the spectrum.

Storage Media

Another basis for definition of A-ID products is the storage media used; paper and ink, ferrous oxide on mylar, EEPROM or any variation. Also, some products are more than a storage media and a method of attachment. They incorporate control or data acquisition and processing logic. This is true of RF/DC terminals, RF smart cards and tags. Isn't this confusing?

This is confusing, not only to members of the industry unable to provide GOOD answers, but potential customers trying to find good solutions to their problem.

Is all of this confusion bad? Not necessarily!

As long as a person is confused—whether a potential customer or industry member, and confusion stimulates you to seek information about technologies; medias for storage and communication; possibilities that processing can be in the label—or tag—reader or writer, terminal, node or host, you have not made a mistake. You may be near the discovery of a higher-quality solution—or a new product.

Let's look at a hypothetical case. There are a variety of smart cards on the market today—cards incorporating a keyboard and display. They function as a calculator or personal computer and a credit card. There also are smart cards without keyboards and displays communicating via RF. If combined, you could have a credit card-sized RF terminal. What would it be called? If customers had a need for it, what would they ask for? It is something new, with no precise term in existence.

The products either come into existence primarily in two ways. First, a company goes outside the "business or industry they are in" to combine existing technologies in a new way, or second potential customers describe problems facing them and the attributes of various products that are needed to solve the problem and this description becomes the blue print for a totally new product.

Constructive confusion

In either case, confusion about what exists, how it works, and what are the limits, and what other possibilities exist, stimulate creative thought and results in a new product—as in the hypothetical example—or a new way to apply existing ones, resulting in a higher quality solution for users. We might call the phenomena "constructive confusion."

Here is an actual situation to illustrate. Single Chip Systems has a chip which transmits not only its ID, but when attached to a microprocessor, it will transmit data the micro has acquired from any source—from sensors, via an A to D converter, a bar code reader, etc.—using a broad range of frequencies, including infrared.

What is it? What should it be called? A smart RF/ID chip? An ID and DC chip? It is confusing. It may also be an opportunity.

"What industry are we in?" Maybe we shouldn't care too much about defining things and just stay constructively confused.

Ron Ames is the founder and president of Ames & Associates—a member of Bushnell Consulting Group—and was the first Chairman of the RF/ID Committee of AIM USA.

Opportunities and Strategies
for the Future

If you are as enthusiastic as I am about the RF/ID and Communicating Electronic Documents industry, it is probably time for you to start thinking about strategies for the future. You should have a strategy for your company—if you are in one—and for you personally, whether you have a professional interest or not.

If you are not in a company you may be an educator, an investor, or maybe a consumer who has an interest in what technology can do to make your life better.

If this is your first pass through the book you may want to finish it and reread it before you start thinking seriously about your strategy.

Before you implement a strategy you need to understand the opportunities in greater depth than they can be presented here. However, you can determine, in general, the classes of applications that are of interest and the role that you are best suited to play.

In the next section I will review applications that I believe represent the best opportunities for the technology and some issues facing the industry.

6.1 Opportunities and Challenges

for the Radio Frequency Identification Industry

Ron Ames—© April, 1989

The Radio Frequency Identification (RF/ID) Industry has been enjoying growth at or above the rate of the Automatic Identification Industry as a whole, however it still has not deeply penetrated most of the applications which offer the greatest potential.

I believe it is time for everyone having a stake in the industry to take time off from their daily treadmills and take an objective look at what must be done, and what they can do, in order to achieve that potential.

Those having a stake in the industry are a large community. In my opinion, they include all of those who benefit from it; by using the products, by investing in a growth industry, by providing materials and services to the industry; as well as those who collect a salary, a commission, or a fee directly from the industry.

Here are seven key areas which must be addressed in order for this industry to reach its full potential.

1. TECHNOLOGY
2. STANDARDIZATION
3. PRODUCTS
4. REGULATION
5. MARKETING
6. INVESTMENT
7. TIME

Before I address these, I will identify some applications which, I believe, represent large potential markets and explore what is required to actually create them. I hope these already are, or soon will be, reflected in well developed and funded business plans.

Tires

This may seem like a strange place for an RF/ID tag. I can see some of you imagining a lump the size and shape of a pack of cigarettes, covered with tread going ker-thump, ker-thump, as the wheel turns. Obviously this wouldn't work.

What is needed is a tiny silicon chip, integrated into the tire, possibly using the wire bead and/or the steel belts as antennas. It would be used to track the tire through the factory to where it is loaded for shipment, through the distribution network to the point of sale, and later to validate warranty claims and to provide quality assurance feedback.

It should have a cost ranging from less than $0.50 for automobile tires to as much as $5.00 for truck, airplane or large industrial tires. The market is hundreds of millions of units per year.

Automobiles

The need to identify this very important piece of personal property occurs many times during its life. It begins in the factory and continues through the distribution channels to the dealership where it is used for inventory control. This is vitally important to the lending institutions loaning the money to buy the inventory. Since the vehicles are on display on open lots, RF/ID would be a major factor in theft prevention and also in achieving higher rates of stolen vehicle recovery.

The tags should not be transferable between vehicles. Rather, any attempt to remove them should destroy them. Title transfers would be contingent on reading the tags and correlating the information with a data base.

Law enforcement agencies would be able to read them, preferably at some distance. Vehicle registration, taxation, use fees, tolls, facility access privileges, vehicle warranty, and service records would all use this form of identification.

The market is not limited to new vehicles. All existing vehicles— hundreds of millions in the U.S. alone—could be retrofitted with RF/ID tags in the form of electronic license plates. The cost, which could be absorbed by the market, is in the $5.00 to $25.00 range. —The automotive industry says less than $1.00 but I disagree.—

Personal Data

The use of a smart card as a personal credential and transaction processor has been widely forecast. However, I believe many people still have not grasped the full range of applications that these devices will ultimately solve. Most people still think of them as a one-for-one replacement for the current, broadly-used credit card. Their current cost, level of convenience, limited capacity and life do limit them to little more than this at present. As a result their cost/benefit ratio has not been sufficiently attractive to displace the older mag-stripe cards.

When they have a capacity of 128KB and greater, they can begin serving as a single integrated personal data base. They can replace all of your credit cards and all of your other credentials. This includes those that you carry with you, and the others you currently leave home because they exceed the level of inconvenience you will routinely tolerate.

Medical records that could save your life in an emergency, military service records, financial accounts, drivers' license, employment history, access privileges to the facility at work and the computer, copier or other equipment you use there, athletic or social club memberships, and security clearances are a few examples of what can be stored.

Essentially any information relevant or potentially useful to you becomes a candidate for storage in your personal data base, accessible only by you or with your permission.

The convenience and increased life—five to ten years—made possible by RF, coupled with a cost of 10 or 20 dollars, will make these attractive to a great many people in the world, possibly 500 million, who are currently paying many times this amount in obvious and hidden costs.

A smaller subset of personal information is the ID card or badge which is currently being used primarily for facility access control. Even though it currently represents one of the largest uses of RF/ID it still represents only a small fraction of the total quantity of card and badge credentials being sold. If the incremental cost of this technology to the user can be reduced to 100% or double the cost of a card using other technologies such as Wiegand wire, Barium ferrite or mag stripe, the volume of RF/ID cards will increase several fold over today's sales.

Animals

Although they were one of the first entities identified by RF/ID tags, the potential for identification of all pets, laboratory animals, farm and ranch animals is still an unachieved dream. Hundreds of thousands of tags have been sold in the last ten years, but the world population of animals in the categories mentioned are counted in hundreds of millions.

Some of these animals can be identified satisfactorily with external tags, but most are better served by injecting tiny, harmless tags under the skin or into their muscle tissue.

The cost of the tags—currently $8 to $10 rather than $3 to $5—and readers are only part of the reason this market has not reached its potential. More important is the frustrating lack of easily accessible

Opportunities and Strategies for the Future

registry data bases and distribution channels to provide the marketing, delivery and support this huge market needs.

Medical

The storage of the ID and other pertinent information in tags attached to blood bags, laboratory test samples, medication, and patients, recording prescribed procedures and in some cases actually controlling their administration, will be a terrific boon to the medical and health care field.

This will be the means for improving the quality of care, providing better documentation, and lowering the cost of providing data to insurers and regulatory agencies.

Small, low-cost—$.50 to $3—highly reliable, read-only and read/write tags are needed in very special packaging to satisfy these needs. This market offers a potential of hundreds of millions or perhaps billions of tags.

Baggage

World wide fears of terrorists' bombs are, unfortunately, well founded. In addition to the convenience of having your baggage handled automatically from the sidewalk of one airport to the baggage carousel of your final destination,—and being confident that it will almost always arrive when and where you expect it—RF/ID tagging of your baggage can play a significant role in preventing a terrorist from succeeding in having his bomb be your bags' traveling companion.

They can prevent the bag—containing the bomb—from being loaded if the ticketed passenger who checked it does not get on the plane. Or, if the passenger debarks before the ticketed destination, the bag can be removed.

There have been a few instances of suicide bombings and bombs added the baggage of the unsuspecting, but if the ones which are not of this type could be prevented, we would all feel better when we board a plane.

This application needs a very low cost— <$0.50 —, write-once tag which can be read at a range of 30 to 36 inches in any orientation in a baggage sortation system at least 98 percent of the time. Hundreds of millions of tags per year would be used world-wide.

Cargo

The world-wide use of cargo containers by commercial steamship companies, railroads, trucklines and air carriers continues to grow rapidly, as does internal—industrial or military—use. Together they represent a very large market in which RF/ID far out distances—both in range and overall quality of solution—all of the alternative technologies.

The ability to store From-To addresses, routing and history information, as well as the manifest, and to read the data at a range of 30 feet or more is what the users really want. Cost is of secondary importance because the tag life should be the same as the container life and the frequency of writing is rather low, allowing the tag cost to be amortized over a long period of time. This market is probably more than 100 million tags in total.

Tool Management

Pilot level projects best describe the current activity. Read/write tag capacities of 1000 bytes are adequate and easily obtainable, but the read-range should be 1.5 to 1.75 inches in order for these to be incorporated in tool magazines of current design. Tag costs of $30.00 to $50.00 can be justified without recycling.

Range is the biggest obstacle to significantly penetrating the 100 million tools that would justify this cost. The second largest obstacle is the availability of CNC computer software which incorporates the use of RF/ID tags. Both of these problems are solvable.

Material Handling

While this is conceptually the same as the cargo application, the containers used are typically small, the range required is short and the cost sensitivity is much greater. The packaging must be rugged and insensitivity to orientation is a definite advantage. The capacity required ranges from about 32 bits to several thousand bytes. While the material handling tags are currently not intelligent, they could incorporate a micro in the future.

In many cases, the tags are attached directly to the material, which may be on pallets, or to the pallets themselves. I can't estimate the potential number of tags represented by this application with any accuracy, but I think most will agree it is many million and an acceptable cost would be from $5 to $30 depending on capacity.

Work-in-Process (WIP)

This application differs from the previous one more in respect to the information stored and in the way it is used, than in capacity or packaging. The WIP tags are currently not intelligent. In the future they could incorporate a micro and probably to greater a advantage than in material handling.

The current price range is $30 to $150. This needs to come down to $15 to $50 to achieve significant penetration.

I can't accurately estimate the potential number of tags represented by this application, but I think most will agree it is several million.

Inventory Control

There is a minimum acceptable ratio of tag cost/tagged item value. This is hard to define since there are some items which could use recyclable tags, and others where they would need to be the use-once-and-destroy type. The latter probably would require that the price be 1 percent or less of the value of the tagged item.

I believe a 32-bit flexible, self-adhesive tag costing less than $.50 is possible in volumes of 5 to 10 million per year which should be easily obtainable at that cost. There are potentially billions of valuable items that would justify this cost and would benefit from the inherent advantages of RF/ID over bar code.

A summary of the potential tag sales over a span of 3-5 years which would be possible if all of the items were identified with RF/ID tags is shown in table 6-1.

Table 6-1
**Potential RF/ID Tag, Memory Card
and Smart Card Sales
Next 3 - 5 Years***

The Markets are:	Tag Class*	Low Est. in $B	High Est. in $B
A. Tires	IID	0.1	10.0
B. Autos	IID	1.0	50.0
C. Personal Data	IVD,IVS	5.0	50.0
D. Animals	IID	0.6	20.0
E. Medical	IID, IIID-S	0.25	15.0
F. Baggage	IID, IIID	0.05	2.5
G. Cargo	IIID	0.20	6.0
H. Tool Management	IIID, IIIS IVD, IVS	0.5	5.0
I. Material Handling	IIID, IVD	0.35	3.5
J. Work-in-Process	IIID, IVD	0.15	1.5
K. Inventory Control	IID	0.005	2.5
	Billion	$8.215	$164.3

* 1. Tag Sales, does not include Scanners, Readers, Reader/Writers or other System
     Components
* 2. Tag Class Definitions: II= 8-128 bits, III= 48-512 bits, IV= 256+bits,
     D= Dumb, S= Smart
* 3. Calculated potential market based on object population,
     available market is not calculable from available data.

Summary of number of applications by class:
7-IID   3-IVS   3-IVD   6-IIID   2-IIIS

I have identified eleven markets, most of which are worthy of the name Block Buster and I have explored some of their product needs. There are, I am sure, others which may deserve to be on this list in place of those I have named, but these make the point that RF/ID is potentially a major industry particularly when one adds the value of the reading and writing equipment.

Now lets take a look at what needs to be done to develop these markets to their full potential.

What specifically needs to be done in TECHNOLOGY, in STANDARDIZATION, in REGULATION, in MARKETING, in INVESTMENT and in TIME.

Technology

In general we must be capable of integrating every component of a tag into a single chip, except for the antenna. This means capacitors, zener diodes, crystals or resonators, microprocessors if required, nonvolatile memory, everything!

We must improve the efficiency—use of power—in the tag in every way possible, including reducing feature sizes to 1.25 or 1.5 microns or whatever is cost effective, incorporate the use of ferroelectric memory and capacitors, and even superconductive on-chip interconnection and antennas as soon as this technology is practical. Automated chip-on-board (COB) or Tape Automated Bonding (TAB) assembly is essential to achieve the costs, volumes and quality needed.

Standardization

Many of these markets such as tires, autos, personal data, baggage, cargo, and tool management require standard products and multi-sourcing to develop their full potential. The others would derive significant benefit from standardization.

Regulation

First, regulations need to be uniform world-wide. Currently individual countries' regulations which are different from other countries are significant impediments to achieving world-wide standardization. In fact, today they mandate non-standardization. Secondarily, they should reflect the economic and social benefits to be gained from these products as compared to the other uses competing for the electromagnetic spectrum.

Marketing

These products are a very significant component in the information management, communication and control capability of the world. They need to be marketed by the same organizations who provide the other essential elements of that capability to the world market.

Whether they are produced by the major information system suppliers, are purchased OEM or are obtained in some other fashion, these products need to have the marketing muscle behind them, including advertising and promotion to the market, and the installation, support and customer service after sale, that other computer products and peripherals currently have.

Investment

Investment is needed to accomplish the tasks identified in the previous paragraphs. It is also needed to bring legitimacy and credibility to this technology because investment is a form of assurance to those considering making large purchases—of what may be perceived by them as a "new" technology. It is also an indication that this is an important industry.

Finally, investment in this industry should be considered on its merit, as a high quality investment in an industry which will see dramatic growth in the next five years while providing major productivity improvements to many other important industries.

Time

When we consider that this technology will make major contributions in personal and international security; the cost and quality of medical and health care; the efficiency of transportation; the productivity of work forces, both agricultural and industrial; the speed and control of product distribution and contribute positively to the quality-of-life of people on a global scale, in many ways in addition to those enumerated here, the question of the timing of these activities may be phrased in the following way.

"Shouldn't the development of this industry be considered in setting the priorities of any industrial nation or multinational corporation?" The urgency is obvious.

If these issues are addressed positively and with a high priority, the RF/ID market, in just a few years, will far surpass the size of the bar code industry today, and will improve our daily lives, in many ways.

Companies already have the base technology. The expertise exists to pull them together do the job. The question remaining is this: Will all of those who have a stake in this industry make the commitment to convert this potential into reality?

6.2 New Manufacturers

When you enter a new business, the most important question is "What business are we going to be in?" This answer must be derived from "What market or application needs a solution, that can we offer, in which we will have an an advantage or that is sufficiently superior to other alternatives in enough respects from the viewpoint of potential customers, that we can expect to profitably sell enough product to operate this business and pay an attractive return to the investors of the capital?"

If that question can be answered satisfactorily, you can move on to answering more detailed questions about how this can be accomplished.

In this process new manufacturers of subsystems in this industry may have a choice to make which has not been available to previous entrants in this role.

In times past, a company typically made a decision whether to use off-the-shelf components to make a tag or to have a custom integrated circuit developed. Even if their choice was the first one, they probably had aspirations or even a plan to do the latter as soon as they had sufficient capital to do so.

Now they may be able to choose to use an off-the-shelf RF/ID tag chip or to do a custom chip development. The alternative of using components which were not intended is probably no longer viable except in cases where large amounts of memory are involved.

At least two companies have stated missions of being suppliers of tag chips to the general marketplace. There will probably be others in the future.

This new alternative source of chips designed specifically for RF/ID tags significantly lowers the cost of entry into this marketplace. The cost of developing a custom chip, which might range from $500,000 dollars to as much as several million, does not have to be included in the business plan, or if the same amount of capital is raised, would allow that amount to be redirected into marketing, a greater variety of packaging or other uses.

So one of the early decisions is the same as if one were going to enter the personal computer business, "Can we, and/or should we, use an off-the-shelf-chip or have one custom designed for our products?"

As in any high-tech start-up situation, you need to have a source of the necessary technology, personnel with the necessary skills and motivation, and you need sufficient capital to develop the products and pay the team until enough business can be generated to be self-sustaining. Usually the team is built around individuals who have experience in either the technology or in the application that you intend to serve.

So pull together your team, decide the business you want to be in, choose your technology strategy, polish your business plan, obtain the funding and give it your best effort. If you have done these, admittedly over simplified steps well, you have an excellent chance and I wish you success.

6.3 Current Suppliers

Current suppliers have already decided the business that they are in and have built a company which is having a certain amount of success. The market and technology are always changing, creating new opportunities and challenges. A business has to be aware of the changes that affect it directly, and should also make some effort to be aware of changes that indirectly offer opportunities or may cause problems later.

A current supplier may become aware that the market they serve would be better served with a different sort of product or maybe the same product but at lower risk to the user, or more convenient access to sales and service support from a local distributor. The current supplier may choose not to do anything until someone else precipitates a change, thereby risking the loss of the market. Or they may initiate changes themselves in order to preempt losing business.

Current suppliers should be looking for new markets that can be served by their current products. They should be looking for other needs in their current markets that could be served by new products. And they should be looking for sources of new technology that can enhance their product line or reduce product cost.

Most current suppliers are continuously looking for other markets for their products. Sometimes the new markets which would be interested in the products are culturally so different than those of primary interest to the company they find that it is counterproductive to supply the new market.

Some of the current suppliers are trying to accumulate enough money to develop a new custom chip in order to serve their current markets more completely. If they can find off-the-shelf chips with suitable functionality, they may not have to pay the cost of a custom chip development.

They should look at methods of manufacturing and packaging their chips, tags and readers which are more automated or require less labor.

6.4 OEM's

OEM's in this context would be manufacturers of products which could incorporate RF/ID components into those products. Examples could be an automatic storage and retrieval system which would have the functional part of RF/ID tags imbedded in the plastic of the boxes or totes in which items are stored or machine tool manufacturers which would have reader/writers built into their machines and tags embedded in the tool adapters.

If you are in this category you may have been successful getting a current supplier to redesign products for your use, and then again you may not.

Companies which have products which are very similar in functionality to your needs, are a good place to start. Develop a specification which shows them what you need. They will probably offer solutions which require some changes in your spec. However, accepting minor changes may save you the cost of designing and developing a custom system starting with a clean sheet of paper.

You should also consider companies which supply RF/ID components. Show them your specification. They probably will be able to provide the system design assistance you will need to incorporate their products into yours or they will quote on the development of a system exclusively for your use.

The last alternative is to design and develop a system in-house using your own engineering staff, possibly assisted by consultants or contracted engineering. This can be a good approach if you have analog and digital engineers on staff as well as integrated circuit designers. There is still some risk in this approach. I know of a medical company who spent $4.5 million and still didn't have a manufacturable design.

6.5 System Integrators

The best strategy for system integrators in RF/ID is first of all to get educated about the basic technology, what it can do and how it works.

Next, get to know the companies who claim to serve the same applications that your company does. Learn about their interfaces, documentation, training and support.

Check their reference installations. They may not have any in your field, or geographically near you, but may have some which are similar to yours in a related field. Since this is a relatively new industry, I would be much more concerned about bad references than I would about no references. We all have to start somewhere, sometime. A great company with great products had to have a first customer.

Make sure they understand you, your company, your applications and expectations. There is a lot of money to be made by integrators who become expert in building RF/ID into the solutions they provide.

6.6 End-users

Last, but certainly not least, I will address the end-user. All products are ultimately used by someone. Even products which are only used in the process of manufacturing other products are under the watchful and critical eye of someone. That someone is as much the end-user as someone who uses an RF/ID tag to open the car door or to access the files on a personal computer.

If you understand how RF/ID works and can visualize how it can provide convenience, security, efficiency, and quality to your personal or professional life you can accept or reject the use of this technology when you are offered the choice in products that you buy or that others provide for your use.

You can be even more proactive and suggest to people who try to persuade you to buy or use something, that it might be more acceptable to you if it incorporated the advantages of RF/ID.

This is not likely to get instant results, but if others agree with you, specifications for the next generation products may actually get changed. End users have more clout than they often realize. As my wife says **"Laugh and the world laughs with you, cry and you get what you want!"** You have a right to your strategy, too. Good luck!

RF/ID tags, RF memory cards/tags, and RF smart cards/tags are more than an idea or concept. They exist, they are in production and they are providing documented benefits to their users.

How they will change in order to provide even more benefits for future users we can all only guess. Those of us who have some understanding of the state of the art in a broad way, believe we can see technologies which can be applied to bring significant improvements to this industry.

There are likely many other innovations which are more suitable than the ones we know, just as there are ways to use these devices in ways we haven't yet thought of.

For these reasons I believe this is a fertile field for those who offer innovation, dedication, curiosity, and hard work in an endeavor to make a living from improving the human condition. I hope this collection of thoughts from many people who are dedicated to this objective has served to stimulate and motivate you too.

Other Sources of information

The only other book I am aware of on this subject is the study "Radio Frequency I.D.: Technologies, Markets and Applications" by Richard D. Bushnell Jr. and me published by Cutter Information Corporation.

Seminar programs which are a part of trade shows such as Scan Tech, sponsored by AIM, and ID Expo, sponsored by ID Systems magazine, are good additional sources of information.

The trade press continues to be a good on-going source of articles such as those found here. I urge to add your name to their list of subscribers.

Last, but certainly not least, information is available from the current suppliers of RF/ID products. All you have to do is ask them.

RF/ID Glossary

Term   Definition

Accuracy—The determination of whether an element of data varies from the actual or intended value.

Active Tags—Tags which use batteries as a partial or complete source of power. They are further differentiated by separating them into those with replaceable batteries and those which have the batteries inside a sealed unit or what may be termed unitized active tags.

Addressability—The ability to address bits, bytes, fields, files or other portions of the storage in a tag.

Alignment—The orientation of the tag to the reader in pitch, roll, and yaw.

Alphanumeric—The character set that contains letters, numbers, and usually other characters, such as punctuation marks.

Antenna—Antennas are the conductive elements which radiate, and/or receive energy in the radio frequency spectrum, to and from the tag.

ASCII—The character set and code described in American National Standard for Information Interchange, ANSI X3.4-1977. A standard code of seven bits representing upper and lower case alphabetic, numeric, symbolic and control characters, used for information interchange between data processing systems, communication systems and associated equipment.

Bar Code—An automatic identification technology that encodes information into an array of adjacent varying width parallel rectangular bars and spaces.

Bidirectional—Capable of operating in either of two directions which are the opposite of each other. For example, a tag which can be read or written from either side is bidirectional.

Bit—A single binary digit, 0 or 1 in a single number, is a bit.

Bit error—When a bit is misinterpreted—0 is seen as a 1—, then a bit error has occurred.

Byte—Number of bits representing a character. Typically ASCII characters are represented by 8 or 7 bits.

Capacity—The number of bits or bytes that can be programmed into a tag. This may represent the bits accessible to the user or the total number including those reserved to the manufacturer e.g. parity or control bits.

Capture Window/Field—Region of the scanner field in which a tag will operate.

Closed Systems—A system in which relevant data regarding the attributes of the object is stored in a common data base, accessible via data link by referencing the individual ID code. It usually refers to a system under the control of a single owner or authority.

Code Plate—See Tag

Controller—See Multiplexer

Electromagnetic coupling—Systems which use a magnetic field as a means of transferring data or power are said to use electromagnetic coupling.

Electrostatic coupling—Systems which use the the inducing of a voltage on a plate as a means of transferring data or power are said to use electrostatic coupling.

Electronic label—See Tag

Error—Any operation or data which is not in accord with the design or input to the system.

Error Correcting Code (ECC)—Supplemental bits in a data transfer used in conjunction with a polynomial algorithm, in order to compute the value of missing or erroneous data bits—e.g. for a 32 bit data transmission, 7 additional bits are required.

Error Correcting Mode—Mode of data communication in which missing or erroneous bits are automatically corrected.

Error Correcting Protocol—The rules by which the error correcting mode operates.

Error Management—Techniques used to ensure that only correct information is presented to the user of the system.

Error Rate—The number of errors per number of transactions.

Exciter—The electronics which drive an antenna are called the exciter or transmitter. Together with the antenna they are called a scanner.

Expansion Port—A plug accessing additional I/O capability on a computer or peripheral device.

Factory Programming—The programming of information into a tag occurring as part of the manufacturing process resulting in a read only tag.

Field Programming—Programming information into the tags may occur after the tag has been shipped from the manufacturer to an OEM customer or end user or in some cases to the manufacturers distribution locations. field programming usually occurs before the tag is installed on the object to be identified. This approach enables the introduction of data relevant to the specifics of the application into the tag at any time, however the tag would typically have to be removed from it's object. In some cases, change or duplication of all data in the tag is possible. In other cases, some portion is reserved for factory programming. This might include a unique tag serial number, for example.

Field Protection—The ability limit the operations which can be performed on portions or fields of the data stored in a tag.

Flat Panel Antenna—Flat, conductive sheet antennas, usually made of metal plate or foil.

Frequency—The number of times a signal executes a complete excursion through its maximum and minimum values and returns to the same value—e.g. cycles.

I.D. Filter—Software that compares a newly read ID with those in a data base or set.

Inductive Coupling—Systems which use the the inducing of a current in a coil as a means of transferring data or power are said to use inductive coupling.

In-Use Programming—Many applications require that new data or revisions to data already in the tag, be entered into the tag, while it remains attached to its object. The ability to read from and write data to the tag while attached to its object is called in-use programming. Tags and systems with this capability are called read/write tags and systems.

Interrogator—See Reader and Programmer

Life—Functional period within which no maintenance, adjustment or repair is to be reasonably expected.

Memory Cards—a read/write or reprogrammable tag in credit card size

Memory Modules—a read/write or reprogrammable tag

Misread—A condition that exists when the data presented by the reader is different from the corresponding data in the tag.

Mobile Inventory Vehicle—Vehicle equipped with a system for locating tagged vehicles, containers, or objects for the propose of inventory control.

Modulation—The method of modulating,or altering the carriers in order to carry the encoded information, are quite varied. They include amplitude modulation (AM), phase modulation (PM), frequency modulation (FM), frequency shift keyed (FSK), pulse position (PPM), pulse duration (PDM) and continuous wave (CW). In some cases, different modulating techniques are used in each direction—to and from the Tag.

Modulation, amplitude (AM)—Data is contained in changes in amplitude of the carrier.

Modulation, phase (PM)—Data is contained in the changes in phase of the carrier.

Modulation, frequency (FM)—Data is contained in the changes in frequency of the carrier.

Modulation, frequency shift keyed (FSK)—Data is contained in the changes between two frequencies of carrier.

Modulation, pulse duration (PDM)—Data is contained in the duration of pulses.

Modulation, pulse position (PPM)—Data is contained in the position of pulses relative to a reference point.

Modulation, pulse duration (PDM)—Data is contained in the duration relative to a reference duration.

Modulation, continuous wave (CW)—Data is contained in a carrier which is switched on and off.

Multiplexer (multiplexor)—A device which supports multiple scanners or antennas by checking each in accordance with some scheduling scheme which may be either round robin or priority based. This reduces the total amount of electronics in the system at the expense of having all scanners being "blind" part of the time. These devices are called multiplexers or multichannel readers or just controllers.

Nominal—The value at which a system is designed assure optimal operation. Tolerances consider the "normal" deviation of variable factors.

Nominal Range—The range at which a system can assure reliable operation, considering the normal variability of the environment in which it is used.

MTBF—Mean Time Between Failure is the manufacturers estimated or calculated time in which a user may expect a device to operate before a failure is experienced.

MTTR—Mean Time To Repair.

Omnidirectional—Capability of a tag to operate in any orientation.

Open Systems—Application in which reader/writers do not have access to a common data base.

Orientation—Alignment of the tag with respect to the scanner, measured in pitch, roll, and yaw.

Orientation Sensitivity—The degree range is decreased by nonoptimal orientation.

Passive Tags—Passive tags contain no internal power source. They are externally powered and typically derive their power from the carrier signal radiated from the scanner.

Port Concentrator—A device that accepts the output from a number of communications interfaces and introduces them into a communications network.

Power Levels—Levels of power radiated from a scanner or tag, usually measured in volts/meter.

Programming—Adding or altering data in a tag.

Programmability—In order to be identifiers of specific objects, tags must at some point have their identity and/or other data entered into them. This capability is called programmability.

Programmer—Some tags which can have their contents changed by a set of electronics in close proximity or in electrical contact with it. Those electronics and their packaging are called a programmer.

Projected Life—This is defined in terms of number of read and/or write cycles, or in active tags this may include shelf life.

Proximity sensor—A device that detects and signals the presence of a selected object at or near the sensors location.

RF/DC—Systems which communicate over a radio link between a host computer and a data source e.g. Keyboards, data terminals,readers for OCR, Bar Codes, Mag Stripes, RF/ID etc. RF/DC enhances the capabilities of Automatic ID systems by providing the capabilities of hard wired data communications without the physical restrictions of interconnecting wires.

RF/ID—Systems that read or write data to RF tags that are present in a radio frequency field projected from RF reading/writing equipment. Data may be contained in one (1) or more bits for the purpose of providing identification and other information relevant to the object to which the tag is attached.It incorporates the use of electromagnetic, or electrostatic coupling in the radio frequency portion of the spectrum to communicate to or from a tag through a variety of modulation and encodation schemes.

RF/AIS—Radio Frequency Automatic Identification Systems

Range—The distance at which successful reading and/or writing can be accomplished.

Read—The decoding, extraction and presentation of data from formatting, control and error management bits sent from a tag.

Read Only—See Factory Programming

Readability—The ability to extract data under less than optimal conditions.

Read Rate—The maximum rate at which data can be read from a tag expressed in bits or bytes per second.

Read/Write—Many applications require that new data or revisions to data already in the Tag, be entered into the Tag, while it remains attached to its object. Tags with this capability are said to be reprogrammable and are called read/write tags, memory cards or memory modules .

Reader—The device containing the digital electronics which extract and separate the information from the format definition and error management bits. The digital electronics perform the actual reading function. These read electronics may also interface to an integral display and/or provide a parallel or serial communications interface to a host computer or industrial controller.

Reader/Writer—The set of electronics can which change the contents of tags while they remain attached to their object, are called a reader/writer.—See also Reader.

Reprogrammable—Many applications require that new data or revisions to data already in the tag, be entered into the tag, while it remains attached to its object. The ability to read from and write data to the tag while attached to its object is called in-use programming. Tags with this capability are said to be reprogrammable and are called read/write tags, memory cards or memory modules .

SAW—Surface Acoustic Wave. A technology in which radio frequency signals are converted to acoustic signals in a piezoelectric crystalline material. Variations in phase shift in the reflected signal can be used to provide a unique identity.

Signalling technique—A complete description of the modulation, encodation, protocol, and sequences required to communicate between two elements of a system.

Scanner—The antenna/s, transmitter—or exciter—and receiver electronics integrated in a single package called the scanner. They may be combined with additional digital electronics including a microprocessor in a package called a reader.

Sensor—A device that responds to a physical stimulus and produces an electronic signal. See Scanner.

Separation—Operational distance between two tags.

Speed—The rate at which something occurs.

Tag—The transmitter/receiver pair or transceiver plus the information storage mechanism attached to the object is referred to as the tag, transponder, electronic label, code plate and various other terms. Although transponder is technically the most accurate, the most common term and the one preferred by the Automatic Identification Manufacturers is tag.

Transponder—See Tag.

Verify—To assure that the intended operation was correctly performed.

Write—The transfer of data to a tag, the tags internal operation of storing the data and it may include reading the data in order to verify the operation.

Write Rate—The rate at which information is transferred to a tag, written into the tags memory and verified as being correct. It is quantified as the average number of bits or bytes per second in which the complete transaction can be performed.

Index

(cut here)

- -

Please enter my name in a drawing for the new edition of:

RADIO FREQUENCY ID; TECHNOLOGIES, MARKETS, AND APPLICATIONS

Name:_____

Title:_____

Company Name:_____

Address:_____

_____

Phone Number:_____

- -

(cut here)

Send To:
    Ames & Associates
    2854 S. Wheeling Way
    Aurora, CO 80014